AF646574

ENCYCLOPAEDIA OF ENVIRONMENTAL BIOTECHNOLOGY

ENCYCLOPAEDIA OF ENVIRONMENTAL BIOTECHNOLOGY

Vol. 4

Shahzad Ahmad

ANMOL PUBLICATIONS PVT. LTD.
NEW DELHI - 110 002 (INDIA)

ANMOL PUBLICATIONS PVT. LTD.
H.O.: 4374/4B, Ansari Road, Darya Ganj,
New Delhi-110 002 (India)
Ph.: 23278000, 23261597
B.O.: No. 1015, Ist Main Road, BSK IIIrd Stage
IIIrd Phase, IIIrd Block,
Bangalore - 560 085 (India)
Visit us at: www.anmolpublications.com

Encyclopaedia of Environmental Biotechnology

© Author

First Published, 2008
ISBN 978-81-261-3539-4 (Set)

PRINTED IN INDIA

Printed at Mehra Offset Press, Delhi.

Contents

Preface ix

1. **Ecological Factors** 1
Various Ecosystems • Climatic Factors

2. **Endocrine Disruptor Chemicals** 51
Fish Reproductive Health • Fish as a Monitor • Organochlorine Pesticides • Industrial Chemicals • Herbicides • Immune System • Testicular Structure • Gamete Viability • Future Needs

3. **Cryoconservation of Fish Gametes** 71
Embryonic Cells • Ice Formation in Solution • Super Cooling and Seeding • Thawing and Fertilization • Oocytes • Somatic Cells • Germ Cell Transplantation

4. **Toxicology and Biomedical Research** 85
Transgenic Fish • Transgenic Fish Models • Transgenic Fish Mutation Models • Choice of Species • Gene Transfer • Cancer Study • Apoptosis Study • Animal Proteomics • Tool in Mosquito Control • Tool in Ontogeny • Biotechnological Research

5. **Environmental Biotechnology and Agriculture** 129
Biofertilizer • Impact on Agriculture • Biofertilizers, Biopesticides • Guidelines for Evaluating Products • Specific Product Groups

6. **Biotechnology and Sustainable Development** 145
Impact of Terminator Technology • Environmental Effects • Agronomic Problems • Stakeholder Perception • Media and Information • Academia • Regulations • Dimensions of Risk • Professional Ethics • Future of Bioethics • Policy Directions • Coconut Water • Light Microscopy • Creation of Constructs • Tobacco Transgenic Plants • Pollen Germination in Vitro • Computer Analysis • Statistical Analysis • Materials and Methods • Kinetic Studies • Desorption of Copper • Bioreactor Cultivations • Mutagenesis • Selection of Mutants • Analytical Methods • Optimisation of Media • Statistical Experimental Design • Mapping Human Variations • Coffee Tissue Culture • Significant Role of Biotechnology • Agriculture and Nutrition • Plant Production • Animal Production • Food Technology • Other Applications • Medicine • Genetic Engineering • Effects of Transgenic Plants • Risks of Biotechnology • Inherent Risks • Transcendent Risks • International Imbalances • National Imbalances • Raw Material Substitutions • Role of Private and Public Sector • Bio and Food-safety • Industrial Biotechnology

7. **New Marine Biotechnology** 291
Options and Opportunities • Concept of Marine Biotechnology • Major Applications and Benefits • Enzymes and Biocatalysts

• Biomolecular Materials • Biosensors and Bioindicators • Biopesticides • Biorefining • Bioremediation • Global Climate Change

Bibliography 309

Index 317

Preface

Environmental biotechnology is the applied technology for the study of natural environment. Environmental biotechnology could also imply that one tries to harness biological process for commercial use and exploitation. Environmental biotechnology, by contrast, deals readily and apparently with dramatic topics, and though their importance, albeit different, may be every bit as great as their direct relevance is quite readily appreciated by the bulk of population. Cleaning up contamination and dealing rationally with waste is of course, in everybody's best interests, but for most people, this is like simply addressing a problem. Some of the potentially beneficial uses of biological engineering, which touch the lives of the majority of people, however directly involve much simpler approaches. In many respects, there is a logical link between the aims of environmental biotechnology.

Environmental biotechnology is fundamentally rooted in wastes, in its various guises, typically being concerned with the remediation of the contamination caused by previous use and the impact reduction of current activity or the control of pollution. Thus, the principal aim of this stream are the manufacture of products in environmentally harmonious manner, which allow for the minimisation of harmful solids, liquids or gaseous outputs or clean up the residual effects of earlier human occupations. Environmental biotechnologists may enhance or optimise conditions for existing biological systems to make their activities happen faster or more efficiently or they resort to some form of alteration to bring about the desired results. The variety of organism, which may play a part in environmental applications of biotechnology is wide and ranging from microbes through trees and all are utilised on one of the same three fundamental bases.

Accordingly, the range of businesses to which environmental biotechnology has potential relevance is almost limitless. All commercial operations generate wastes in one form or the other. Additionally, biotechnology can be of further commercial significance by converting low-cost organic feed stocks into high-value products.

Biotechnology based methods are seen as essential to help meet European Union (EU) targets for biowaste diversion and reductions in pollutants. Across the world, the existing regulations on environmental pollution are required to be rigorously enforced, with more stringent compliance standards implemented. All of this is expected to stimulate the sale of biotechnology based environmental processing methods significantly and in particular, the global market share is projected to grow faster than the general biotic sector trends.

This exhaustive work has a perfect treatment with environmental biotechnology in an objective manner. Expectedly, this work would be capable to create a sort of awareness among the people and make the study of environmental biotechnology easier and more meaningful.

Editor

1

Ecological Factors

Idea of Ecosystem

Various kinds of life-supporting systems like the forests, grasslands, oceans, lakes, rivers, mountains, deserts and estuaries show wide variations in their structural composition and functions. However, they all are alike in the fact that they consist of living entities interacting with their surroundings exchanging matter and energy.

How do these different units like a hot desert, a dense evergreen forest, the Antarctic Sea or a shallow pond differ in the type of their flora and fauna, how do they derive their energy and nutrients to live together, how do they influence each other and regulate their stability are the questions that are answered by ecology.

The term ecology was coined by Earnst Haeckel in 1869. It is derived from the Greek words *Oikos-* home + *logos-* study. So ecology deals with the study of organisms in their natural home interacting with their surroundings. The surroundings or environment consists of other living organisms (biotic) and physical (abiotic) components. Modern ecologists believe that an adequate definition of ecology must specify some unit of study and one such basic unit described by Tansley (1935) was ecosystem. An ecosystem is a group of biotic communities of species interacting with one another and with their non-living environment exchanging energy and matter. Now ecology is often defined as "the study of ecosystems".

An ecosystem is an integrated unit consisting of interacting plants, animals and microorganisms whose survival depends upon the maintenance and regulation of their biotic and abiotic structures and functions.

The ecosystem is thus, a unit or a system which is composed of a number of subunits, that are all directly or indirectly linked with each other. They may be freely exchanging energy and matter from outside—an open ecosystem or may be isolated from outside—a closed ecosystem.

Features of Ecosystem

Ecosystems show large variations in their size, structure, composition, etc. However, all the ecosystems are characterized by certain basic structural and functional features which are common.

Structural Features : Composition and organization of biological communities and abiotic components constitute the structure of an ecosystem.

Biotic Structure : The plants, animals and microorganisms present in an ecosystem form the biotic component. These organisms have different nutritional behaviour and status in the ecosystems and are accordingly known as Producers or Consumers, based on how do they get their food.

(a) *Producers:* They are mainly the green plants, which can synthesize their food themselves by making use of carbon dioxide present in the air and water in the presence of sunlight by involving chlorophyll, the green pigment present in the leaves, through the process of photosynthesis. They are also known as photo autotrophs (auto=self; troph=food, photo=light).

There are some microorganisms also which can produce organic matter to some extent through oxidation of certain chemicals in the absence of sunlight. They are known as chemosynthetic organisms or chemo-autotrophs. For instance in the ocean depths, where there is no sunlight, chemoautotrophic sulphur bacteria make use of the heat generated by the decay of radioactive elements present in the earth's core and released in ocean's depths. They use this heat to convert dissolved hydrogen sulphide (H_2S) and carbon dioxide (CO_2) into organic compounds.

(b) *Consumers:* All organisms which get their organic food by feeding upon other organisms are called consumers, which are of the following types:

(i) *Herbivores (plant eaters):* They feed directly on producers and hence also known as primary consumers, e.g. rabbit, insect, man.

(ii) *Carnivores (meat eaters):* They feed on other consumers. If they feed on herbivores they are called secondary consumers (e.g. frog) and if they feed on other carnivores (snake, big fish, etc.) they are known as tertiary carnivores/consumers.

(iii) *Omnivores:* They feed on both plants and animals, e.g. humans, rat, fox, many birds.

(iv) *Detritivores (detritus feeders or saprotrophs):* They feed on the parts of dead organisms, wastes of living organisms, their cast-offs and partially decomposed matter e.g. beetles, termites, ants, crabs, earthworms, etc.

(c) *Decomposers:* They derive their nutrition by breaking down the complex organic molecules to simpler organic compounds and ultimately into inorganic nutrients. Various bacteria and fungi are decomposers.

In all the ecosystems, this biotic structure prevails. However, in some, it is the primary producers which predominate (e.g. in forests, agro-eco systems) while in others the decomposers predominate (e.g. deep ocean).

Abiotic Structure: The physical and chemical components of an ecosystem constitute its abiotic structure. It includes climatic factors, edaphic (soil) factors, geographical factors, energy, nutrients and toxic substances.

(a) *Physical Factors:* The sunlight and shade, intensity of solar flux, duration of sun hours, average temperature, maximum-minimum temperature, annual rainfall, wind, latitude and altitude, soil type, water availability, water currents, etc. are some of the important physical features which have a strong influence on the ecosystem.

We can clearly see the striking differences in solar flux, temperature and precipitation (rainfall, snow, etc.) pattern in a desert ecosystem, in a tropical rainforest and in tundra ecosystem.

(b) *Chemical Factors:* Availability of major essential nutrients like carbon, nitrogen, phosphorus, potassium, hydrogen, oxygen and sulphur, level of toxic substances, salts causing salinity and various organic substances present in the soil or water largely influence the functioning of the ecosystem.

All the biotic components of an ecosystem are influenced by the abiotic components and vice versa, and they are linked together through energy flow and matter cycling .

Functional Characters

Every ecosystem performs under natural conditions in a systematic way. It receives energy from the sun and passes it on through various biotic components and in fact, all life depends upon this flow of energy. Besides energy, various nutrients and water are also required for life processes which are exchanged by the biotic components within themselves and with their abiotic components within or outside the ecosystem. The biotic components also regulate themselves in a very systematic manner and show mechanisms to encounter some degree of environmental stress. The major functional attributes of an ecosystems are as follows:

(i) Food chain, food webs and trophic structure

(ii) Energy flow

(iii) Cycling of nutrients (Biogeochemical cycles)

(iv) Primary and Secondary production

(v) Ecosystem development and regulation

Trophic Structure **:** The structure and functions of ecosystems are very closely related and influence each other so intimately that they need to be studied together. The flow of energy is mediated through a series of feeding relationships in a definite sequence or pattern which is known as food chain. Nutrients too move along the food chain.

The producers and consumers are arranged in the ecosystem in a definite manner and their interaction along with population size are expressed together as trophic structure. Each food level is known as trophic level and the amount of living matter at each trophic level at a given time is known as standing crop or standing Bio-mass.

Before we study about energy flow or nutrient cycling, we must learn about the food-chains, that provide the path through which the flow of energy and matter take place in ecosystem.

Chain of Food

The sequence of eating and being eaten in an ecosystem is known as food chain. All organisms, living or dead, are potential food for some other organism and thus, there is essentially no waste in the functioning of a natural ecosystem. A caterpillar eats a plant leaf, a sparrow eats the caterpillar, a cat or a hawk eats the sparrow and when they all die, they are all consumed by microorganisms like bacteria or fungi (decomposers) which break down the organic matter and convert it into simple inorganic substances that can again be used by the plants—the primary producers.

Some common examples of simple food chains are:

- Grass → grasshopper → Frog → Snake → Hawk (Grassland ecosystem)
- Phytoplanktons → water fleas → small fish → Tuna (Pond ecosystem)
- Lichens → reindeer → Man (Arctic tundra)

Each organism in the ecosystem is assigned a feeding level or trophic level depending on its nutritional status. Thus, in the grassland food chain, grasshopper occupies the 1st trophic level, frog the IInd and snake and hawk occupy the IIIrd and the IVth trophic levels, respectively. The decomposers consume the dead matter of all these trophic levels. In nature, we come across two major types of food chains:

I. *Grazing Food Chain:* It starts with green plants (primary producers) and culminates in carnivores. All the examples cited above show this type of food chain. Another example could be Grass → Rabbit → Fox

II. *Detritus Food Chain:* It starts with dead organic matter which the detritivores and decomposers consume. Partially decomposed dead organic matter and even the decomposers are consumed by detritivores and their predators. An example of the detritus food chain is seen in a mangrove (estuary).

Here, a large quantity of leaf material falls in the form of litter into the water. The leaf fragments are eaten by saprotrophs. (Saprotrophs are those organisms which feed on dead organic matter). These fallen leaves are colonized by small algae, which are also consumed by the saprotrophs or detritivores consisting of crabs, mollusks, shrimps, insect larvae, nematodes and fishes. The detritivores are eaten by small carnivorous fishes, which is turn are eaten by large carnivorous fishes. Leaf litter → algae → crabs → small carnivorous fish → large carnivorous fish (mangrove ecosystem).

Dead organic matter → fungi → bacteria (forest ecosystem).

Thus the grazing food chain derives its energy basically from plant energy while in the detritus food chain it is obtained primarily from plant Bio-mass, secondarily from microbial Bio-mass and tertiarily from carnivores. Both the food chains occur together in natural ecosystems, but grazing food chain usually predominates.

Food Web : Food chains in ecosystems are rarely found to operate as isolated linear sequences. Rather, they are found to be interconnected and usually form a complex network with several linkages and are known as food webs. Thus, food web is a network of food chains where different types of organisms are connected at different trophic levels, so that there are a number of options of eating and being eaten at each trophic level.

A food-web in the unique Antarctic Ecosystem is representing the total ecosystem including the Antarctic sea and the Continental land. The land does not show any higher life forms of plants. The only species are that of some algae, lichens and mosses. The animals include penguins and snow petrel which depend upon the aquatic chain for their food energy.

In a tropical region, on the other hand, the ecosystems are much more complex. They have a rich species diversity and therefore, the food webs are much more complex.

Why nature has evolved food webs in ecosystems instead of simple linear food chains? This is because food webs give greater stability to the ecosystem. In a linear food chain, if one species becomes extinct or one species suffers, the species in the subsequent trophic levels are also affected. In a food web, on the other hand, there are a number of options available at each trophic level. So if one species is affected, it does not affect other trophic levels so seriously.

Just consider the simple food chains of arctic tundra ecosystem: Cladonia → Reindeer → Man Grass → Caribou → Wolf

If due to some stress, the population of reindeer or Caribou falls, it will leave little option for man or wolf to eat from the ecosystem. Had there been more biodiversity, it would have led to complex food web giving the ecosystem more stability.

Significance of Food Chains and Food Webs : Food chains and food webs play a very significant role in the ecosystem because the two most important functions of energy flow and nutrient cycling take place through them.

The food chains also help in maintaining and regulating the population size of different animals and thus, help maintain the ecological balance.

Food chains show a unique property of biological magnification of some chemicals. There are several pesticides, heavy metals and other chemicals which are non-biodegradable in nature. Such chemicals are not decomposed by microorganisms and they keep on passing from one trophic level to another. And, at each successive trophic level, they keep on increasing in concentration. This phenomenon is known as biomagnification or biological magnification.

Pyramids in Ecological System : Graphic representation of trophic structure and function of an ecosystem, starting with producers at the base and successive trophic levels forming the apex is knows as an ecological pyramid. Ecological pyramids are of three types:

Pyramid of Numbers: It represents the number of individual organisms at each trophic level. We may have upright or inverted pyramid of numbers, depending upon the type of ecosystem and food chain. A grassland ecosystem and a pond ecosystem can have upright pyramid of numbers. The producers in the grasslands are grasses and that in a pond are phytoplanktons (algae, etc.), which are small in size and very large in number. So the producers form a broad base. The herbivores in a grassland are insects while tertiary carnivores are hawks or other birds which are gradually less and less in number and hence the pyramid apex becomes gradually narrower forming an upright pyramid. Similar is the case with the herbivores, carnivores and top carnivores in pond which decrease in number at higher trophic levels.

In a forest ecosystem, big trees are the producers, which are less in number and hence form a narrow base. A larger number of herbivores including birds, insects and several species of animals feed upon the trees (on leaves, fruits, flowers, bark, etc.) and form a much broader middle level. The secondary consumers like fox, snakes, lizards, etc. are less in number than herbivores while top carnivores like lion, tiger, etc. are still smaller in number. So the pyramid is narrow on both sides and broader in the middle.

Parasitic food chain shows an inverted pyramid of number. The producers like a few big trees harbour fruit, eating birds acting like herbivores which are larger in number. A much higher number of lice, bugs, etc. grow as parasites on these birds while a still greater number of hyperparasites like bugs, fleas and microbes feed upon them, thus making an inverted pyramid.

Pyramid of Bio-mass: It is based upon the total Bio-mass (dry matter) at each trophic level in a food chain. The pyramid of Bio-mass can also be upright or inverted. The pyramid of bio-mass in a forest is upright in contrast to its pyramid of numbers. This is because the producers (trees) accumulate a huge biomass while the consumers' total biomass feeding on them declines at higher trophic levels, resulting in broad base and narrowing top.

The pond ecosystem shows an inverted pyramid of biomass. The total biomass of producers (phytoplanktons) is much less as compared to herbivores (zooplanktons, insects), Carnivores (Small fish) and tertiary carnivores (big fish). Thus the pyramid takes an inverted shape with narrow base and broad apex.

Pyramid of Energy: The amount of energy present at each trophic level is considered for this type of pyramid. Pyramid of energy gives the best representation of the trophic relationships and it is always upright.

At every successive trophic level, there is a huge loss of energy (about 90%) in the form of heat, respiration, etc. Thus, at each next higher level only 10% of the energy passes on. Hence, there is a sharp decline in energy level of each successive trophic level as we move from producers to top carnivores. Therefore, the pyramid of energy is always upright.

Flow of Energy in an Ecosystem : Flow of energy in an ecosystem takes place through the food chain and it is this energy flow which keeps the ecosystem going. The most important feature of this energy flow is that it is unidirectional or oneway flow. Unlike the nutrients (like carbon, nitrogen, phosphorus, etc.) which move in a cyclic manner and are reused by the producers after flowing through the food chain, energy is not reused in the food chain. Also, the flow of energy follows the two laws of Thermodynamics.

First law of Thermodynamics states that energy can neither be created nor be destroyed but it can be transformed from one form to another. The solar energy captured by the green plants (producers) gets converted into biochemical energy of plants and later into that of consumers.

Dud law of Thermodynamics states that energy dissipates as it is used or in other words, its gets converted from a more concentrated to dispersed form. As energy flows through the food chain, there occurs dissipation of energy at every trophic level. The loss of energy takes place through respiration, loss of energy in locomotion, running, hunting and other activities. At every level there is about 90% loss of energy and the energy transferred from one trophic level to the other is only about 10%.

Energy Flow Models: The flow of energy through various trophic levels in an ecosystem can be explained with the help of various energy flow models.

(a) *Universal Energy Flow Model:* Energy flow through an ecosystem was explained by E.P. Odum as the universal energy flow model. As the flow of energy takes place, there is a gradual loss of energy at every level, thereby resulting in less energy available at next trophic level as indicated by narrower pipes (energy flow) and smaller boxes (stored energy in biomass). The loss of energy is mainly the energy not utilized (NU). This is the energy lost in locomotion, excretion, etc. or it is the energy lost in respiration (R) which is for maintenance. The rest of the energy is used for production (P).

(b) *Single Channel Energy Flow Model:* The flow of energy takes place in an unidirectional manner through a single channel of green plants or producers to herbivores and carnivores.

(c) *Double Channel or Y-shaped Energy Flow Model:* In nature, both grazing food chain and detritus food chain operate in the same ecosystem. However, sometimes it is the grazing food chain which predominates. It happens in marine ecosystem where primary production in the open sea is limited and a major portion of it is eaten by herbivorous marine animals. Therefore, very little primary production is left to be passed on to the dead or detritus compartment. On the other hand, in a forest ecosystem the huge quantity of biomass produced cannot be all consumed by herbivores. Rather, a large proportion of the live biomass enters into detritus (dead) compartment in the form of litter. Hence the detritus food chain is more important there.

The two channel or Y-shaped model of energy flow shows the passage of energy through these two chains, which are separated in time and space.

Function of Nutrient Cycling : Besides energy flow, the other important functional attribute of an ecosystem is nutrient cycling. Nutrients like carbon, nitrogen, sulphur, oxygen, hydrogen, phosphorus, etc. move in circular paths through biotic and abiotic components and are therefore known as biogeochemical cycles.

Water also moves in a cycle, known as hydrological cycle. The nutrients too move through the food chain and ultimately reach the detritus compartment (containing dead organic matter) where various micro-organisms carry out decomposition. Various organically, bound nutrients of dead plants and animals are converted into inorganic substances by microbial decomposition that are readily used up by plants (primary producers) and the cycle starts afresh.

Function of Nitrogen Cycle : Nitrogen is present in the atmosphere as N_2 in large amount (78%) and it is fixed either by the physical process of lightening or biologically by some bacteria and/or cyanobacteria (blue green algae). The nitrogen is taken up by plants and used in metabolism for biosynthesis of aminoacids, proteins, vitamins, etc. and passes through the food chain. After death of the plants and animals, the organic nitrogen in dead tissues is decomposed by several groups of ammonifying and nitrifying bacteria which convert them into ammonia, nitrites and nitrates, which are again used by plants. Some bacteria convert nitrates, into molecular nitrogen or N_2 which is released back into the atmosphere and the cycle goes on.

Function of Carbon Cycle : Sometimes human interferences disturb the normal cycling of such nutrients and create imbalances. For example, nature has a very balanced carbon cycle. Carbon, in the form of carbon dioxide is taken up by green plants as a raw material for photosynthesis, through which a variety of carbohydrates and other organic substances are produced. Through the food chain it moves and ultimately organic carbon present in the dead matter is returned to the atmosphere as carbon dioxide by microorganisms. Respiration by all organisms produces carbon dioxide, while the latter is used up by plants.

In the recent years carbon dioxide levels have increased in the atmosphere due to burning of fossil fuels, etc. which has caused an imbalance in the natural cycle and the world today is facing the serious problem of global warming due to enhanced carbon dioxide emissions.

Function of Phosphorus Cycle : Phosphorous cycle is another important nutrient cycle. The reservoir of phosphorus lies in the rocks, fossils, etc. which is excavated by man for using it as a fertilizer. Farmers use the phosphate fertilizers indiscriminately and as a result excess phosphates are lost as run-off, which causes the problem of eutrophication or overnourishment of lakes leading to algal blooms as already discussed in unit 2. A good proportion of phosphates moving with surface runoff reaches the oceans and are lost into the deep sediments.

Our limited supply of phosphorus lying in the phosphate rocks of this earth are thus overexploited by man and a large part is taken out of the normal cycle due to loss into oceans. So human beings are making the phosphorous cycle acyclic. Sea birds, on the other hand, are playing an important role in phosphorus cycling. They eat sea-fishes which are phosphorus-rich and the droppings or excreta of the birds return the phosphorus on the land. The Guano deposits on the coasts of Peru are very rich sources of phosphorus.

Production in Primary Stage : Primary productivity of an ecosystem is defined as the rate at which radiant energy is converted into organic substances by photosynthesis or chemo-synthesis by the primary producers.

When organic matter is produced by the primary producers (mainly green plants and some microorganisms), some of it is oxidized or burnt inside their body and converted into carbon dioxide which is released during respiration and is accompanied by loss of energy. Respiratory loss of energy is a must, because it is required for the maintenance of the organism. Now, the producers are left with a little less organic matter than what was actually produced by them. This is known as the net primary production (NPP) and the respiratory loss (R) added to it gives the gross primary production (GPP).

Thus, NPP = GPP - R.

Primary production of an ecosystem depends upon the solar radiations, availability of water and nutrients and upon the type of the plants and their chlorophyll content. Below given table shows the average gross primary productivity of some major ecosystems.

Annual Average of Gross Primary Production of Some Major Eco-systems

Ecosystem	*Gross Primary Productivity (K Cal/m²/yr)*
Deserts and Tundra	200
Open Oceans	1,000
Grasslands	2,500
Moist Temperate Forests	8,000
Agro-ecosystems	12,000
Wet Tropical Forests	20,000
Estuaries	20,000

Productivity of tropical forests and estuaries are the highest. This is because tropical forests have abundant rainfall, warm temperature congenial for growth, abundant sunlight and a rich diversity of species. Estuaries get natural energy subsidies in the form of wave currents that bring along with them nutrients required for production.

Deserts, on the other hand, have limitations of adequate water supply while Tundra have very low temperature as limiting factor and hence show low primary production.

Agro-ecosystems get lots of energy subsidies in the form of irrigation water, good quality seeds, fertilizers and pesticides and show a high productivity of 12,000 K Cal/m^2/yr.

Still, it is noteworthy that their productivity is less than that of tropical forests which are not receiving any artificial energy subsidies. Nature itself has designed its species composition, structure, energy capture and flow, and a closed nutrient cycling system that ensures a high primary production of 20,000 K Cal/m^2/yr. Also, the qualitative variety of the primary production is enormous in the tropical forests. This makes it all the more important to conserve our tropical forests.

Production in Secondary Stage : The food synthesized by green plants through photosynthesis is the primary production which is eaten by herbivores. The plant energy is used up for producing organic matter of the herbivores which, in turn, is used up by the carnivores.

The amount of organic matter stored by the herbivores or carnivores (in excess of respiratory loss) is known as secondary production. The energy stored at consumer level for use by the next trophic level is thus defined as secondary production.

Regulating Ecosystem : All ecosystems regulate themselves and maintain themselves under a set of environmental conditions. Any environmental stress tries to disturb the normal ecosystem functions.

However, the ecosystem, by itself, tries to resist the change and maintain itself in equilibrium with the environment due to a property known as homeostasis. Homeostasis is the inherent property of all living systems to resist change. However, the system can show this tolerance or resistance only within a maximum and a minimum range, which is its range of tolerance known as homeostatic plateau.

Within this range, if any stress tries to cause a deviation, then the system has its own mechanisms to counteract these deviations which are known as negative feedback mechanisms. So negative feedback mechanisms are deviation counteracting mechanisms which try to bring the system back to its ideal conditions.

But, if the stress is too high and beyond the range of homeostatic plateau, then another type of mechanisms known as positive feedback mechanisms start operating. These are the deviation accelerating mechanisms. So the positive feedback mechanisms add to the stress conditions and tend to take the system away from the optimal conditions.

Human beings should try to keep the ecosystems within the homeostatic plateau. They should not contribute to positive feedbacks. Otherwise the ecosystems will collapse.

Various Ecosystems

An ecosystem is not static in nature. It is dynamic and changes its structure as well as function with time and quite interestingly, these changes are very orderly and can be predicted. It is observed that one type of a community is totally replaced by another type of community over a period of time and simultaneously several changes also occur. This process is known as ecological succession.

Ecological succession is defined as an orderly process of changes in the community structure and function with time mediated through modifications in the physical environment and ultimately culminating in a stabilized ecosystem known as climax.

The whole sequence of communities which are transitory are known as Seral stages or seres whereas the community establishing first of all in the area is called a *pioneer* community.

Ecological successions starting on different types of areas or substrata are named differently as follows:

(i) *Hydrarch or Hydrosere:* Starting in watery area like pond, swamp, bog

(ii) *Mesarch:* starting in an area of adequate moisture.

(iii) *Xerarch or Xerosere:* Starting in a dry area with little moisture. They can be of the following types:

Lithosere	:	starting on a bare rock
Psammosere	:	starting on sand
Halosere	:	starting on saline soil

Succession Process

The process of succession takes place in a systematic order of sequential steps as follows:

(i) *Nudation:* It is the development of a bare area without any life form. The bare area may be caused due to landslides, volcanic eruption, etc. (topographic factor), or due to drought, glaciers, frost, etc. (Climatic factor), or due to overgrazing, disease outbreak, agricultural/industrial activities (biotic factors).

(ii) *Invasion:* It is the successful establishment of one or more species on a bare area through dispersal or migration, followed by ecesis or establishment. Dispersal of the seeds, spores, etc. is brought about by wind, water, insects or birds. Then the seeds germinate and grow on the land. As growth and reproduction start, these pioneer species increase in number and form groups or aggregations.

(iii) *Competition and Co-action:* As the number of individuals grows there is competition, both inter-specific (between different species) and intra-specific (within the same species), for space, water and nutrition. They influence each other in a number of ways, known as co-action.

(iv) *Reaction:* The living organisms grow, use water and nutrients from the substratum, and in turn, they have a strong influence on the environment which is modified to a large extent and this is known as reaction. The modifications are very often such that they become unsuitable for the existing species and favour some new species, which replace them. Thus, reaction leads to several seral communities.

(v) *Stabilization:* The succession ultimately culminates in a more or less stable community called climax which is in equilibrium with the environment.

The climax community is characterized by maximum biomass and symbiotic (mutually beneficial) linkages between organisms and are maintained quite efficiently per unit of available energy.

Let us consider very briefly two types of succession.

A. Hydrosere (Hydrarch): This type of succession starts in a water body like pond. A number of intermediate stages come and ultimately it culminates in a climax community which is a forest.

The pioneer community consists of phytoplanktons, which are free floating algae, diatoms etc. Gradually these are replaced by rooted-submerged plants followed by rooted-floating plants. Growth of these plants keep on adding organic matter to the substratum by death and decay and thus a layer of soil builds up and shallowing of water takes place. Then Reed swamp (marshy) stage follows in which the plants are partly in water and partly on land. This is followed by a sedge-meadow stage of grasses then by a woodland consisting of shrubs and trees and finally by a forest acting as climax.

B. Xerosere (Xerarch): This type of succession originates on a bare rock, which lacks water and organic matter. Interestingly, here also the climax community is a forest, although the intermediate stages are very different.

The pioneer community here consists of crustose and foliose lichens. These lichens produce some weak acids and help in disintegrating the rock, a process known as weathering. Their growth helps in building up gradually some organic matter, humus and soil. Then comes the community of mosses, followed by herbs, shrubs and finally the forest trees. Throughout this gradual process there is a slow build up of organic matter and water in the substratum.

Thus, succession tends to move towards mesic conditions (moderate condition), irrespective of the fact, whether it started from a dry (Xeric) condition or a moist (hydric) condition and it culminates in a stable climax community, which is usually a forest.

Types of Major Ecosystem : Let us consider types, characteristic features, structure and functions of some major ecosystems.

Ecosystem in Forests

These are the ecosystems having a predominance of trees that are interspersed with a large number of species of herbs, shrubs, climbers, lichens, algae and a wide variety of wild animals and birds. As discussed above forests are found in undisturbed areas receiving moderate to high rainfall and usually occur as stable climax communities.

Depending upon the prevailing climatic conditions forests can be of various types:

(a) *Tropical Rain Forests:* They are evergreen broadleaf forests found near the equator. They are characterized by high temperature, high humidity and high rainfall, all of which favour the growth of trees. All through the year the climate remains more or less uniform. They are the richest in biodiversity. Different forms of life occupy specialized areas (niches) within different layers and spaces of the ecosystem depending upon their needs for food, sunlight, water, nutrient, etc.

We come across different types and layers of plants and animals in the tropical rain forests, e.g. the emergent layer is the topmost layer of the tallest broad-leaf evergreen trees, below which lies the canopy where top branches of shorter trees form an umbrella-like cover. Below this is present the under story of still smaller trees. On the tree trunks some woody climbers are found to grow which are known as Lianas. There are some other plants like orchids which are epiphytes i.e. they are attached to the trunks or branches of big trees and they take up water and nutrients falling from above. The orchids have special type of leaves to capture and hold the water. Some large epiphytes can hold as much as 4 litres of water, equivalent to a small bucket! Thus, these epiphytes almost act like mini-ponds suspended up in the air, in the forest crown. That is the reason why a large variety of birds, insects and animals like monkeys have made their natural homes (habitats) in these forests (Plate II).

The under-storey trees usually receive very dim sunlight. They usually develop dark green leaves with high chlorophyll content so that they can use the diffused sunlight for photosynthesis. The shrub layer receives even less sunlight and the ground layer commonly known as forest floor receives almost no sunlight and is a dark layer. Most of the animals like bats, birds, insects, etc. occupy the bright canopy layer while monkeys, toads, snakes, chameleons etc. keep on moving up and down in sunny and darker layers. Termites, fungi, mushrooms, etc. grow on the ground layer. Warm temperature and high availability of moisture facilitate rapid breakdown (decomposition) of the dropped leaves, twigs, etc. releasing the nutrients rapidly. These nutrients are immediately taken up by the mycorrhizal roots of the trees.

Interestingly, the flowers of forest trees are very large, colourful, fragrant and attractive which helps in pollination by insects, birds, bats, etc. Rafflesia arnoldi, the biggest flower (7 kg weight) is known to smell like rotten meat and attracts flies and beetles which help in its pollination (Plate III).

The Silent Valley in Kerala is the only tropical rain forest lying in India which is the natural habitat for a wide variety of species.

Being the store-house of biodiversity, the forests provide us with an array of commercial goods like timber, fuel wood, drugs, resins, gums, etc. Unfortunately the cutting down of these forests is at an alarming rate. Within the next 30-40 years we are likely to be left with only scattered fragments of such forests, thereby losing the rich biodiversity and the ecological uses of forests, discussed earlier in unit II.

(b) *Tropical Deciduous Forests:* They are found a little away from the equator and are characterized by a warm climate the year round. Rain occurs only during monsoon. A large part of the year remains dry and therefore different types of deciduous trees are found here, which lose their leaves during dry season.

(c) *Tropical Scrub Forests:* They are found in areas where the dry season is even longer. Here there are small deciduous trees and shrubs.

(d) *Temperate Rain Forests:* They are found in temperate areas with adequate rainfall. These are dominated by coniferous trees like pines, firs, redwoods, etc. They also consist of some evergreen broad leaf trees.

(e) *Temperate Deciduous Forests:* They are found in areas with moderate temperatures. There is a marked seasonality with long summers, cold but not too severe winter and abundant rainfall throughout the year. The major trees include broad leaf deciduous trees like oak, hickory, poplar, etc.

(f) *Evergreen Coniferous Forests (Boreal Forests):* They are found just south of arctic tundra. Here winters are long, cold and dry. Sunlight is available for a few hours only. In summer the temperature is mild, sun-shines for long hours but the season is quite short. The major trees include pines, spruce, fir, cedar, etc. which have tiny, needle-shaped leaves having a waxy coating so that they can withstand severe cold and drought. The soil is found to get frozen during winter when few species can survive. The leaves, also known as needles, fall on the forest floor and cover the nutrient poor soil. These soils are acidic and prevent other plants from growing. Species diversity is rather low in these forests.

Ecosystems in Grasslands

Grasslands are dominated by grass species but sometimes also allow the growth of a few trees and shrubs. Rainfall is average but erratic. Limited grazing helps to improve the net primary production of the grasslands but overgrazing leads to degradation of these grasslands resulting in desertification. Three types of grasslands are found to occur in different climatic regions:

(a) *Tropical Grasslands:* They occur near the borders of

tropical rain forests in regions of high average temperature and low to moderate rainfall. In Africa, these are typically known as Savannas, which have tall grasses with scattered shrubs and stunted trees. The Savannas have a wide diversity of animals including zebras, giraffes, gazelle, antelopes, etc. During dry season, fires are quite common. Termite mounds are very common here. The termites gather the detritus (dead organic matter) containing a lot of cellulose and build up a mound. On the top of the mound fungi are found to grow which feed upon this dead matter including cellulose and in turn release methane, a greenhouse gas.

Tropical savannas have a highly efficient system of photosynthesis. Most of the carbon assimilated by them in the form of carbohydrates is in the perennating bulbs, rhizomes, runners, etc. which are present underground. Deliberate burning of these grasslands can release huge quantities of carbon dioxide, another green house gas, responsible for global warming.

(b) *Temperate Grasslands:* They are usually found on flat, gentle sloped hills, winters are very cold but summers are hot and dry. Intense grazing and summer fires do not allow shrubs or trees to grow.

In United States and Canada these grasslands are known as prairies, in South America as Pampas, in Africa as Velds and in central Europe and Asia they are known as Steppes.

Winds keep blowing and evaporation rate is very high. It also favours rapid fires in summer. The soils are quite fertile and therefore, very often these grasslands are cleared for agriculture.

(c) *Polar Grasslands (Arctic Tundra):* They are found in arctic polar region where severe cold and strong, frigid winds along with ice and snow create too harsh a climate for trees to grow. In summers the sun shines almost round the clock and hence several small annual plants grow in the summer. The animals include arctic wolf, weasel, arctic fox, reindeer, etc. A thick layer of ice remains frozen under the soil surface throughout the year and is known as permafrost. In summer, the tundra shows the appearance of shallow lakes, bogs, etc. where mosquitoes, different type of insects and migratory birds appear.

Ecosystems in Deserts

These ecosystems occur in regions where evaporation exceeds precipitation (rainfall, snow etc.). The precipitation is less than 25 cm per year. About 1/3rd of our world's land area is covered by deserts. Deserts have little species diversity and consist of drought-resistant or drought-avoiding plants. The atmosphere is very dry and hence it is a poor insulator. That is why in deserts the soil gets cooled up quickly, making the nights cool. Deserts are of three major types, based on climatic conditions:

(a) Tropical deserts like Sahara and Namib in Africa and Thar desert, Rajasthan, India are the driest of all with only a few species. Wind blown sand dunes are very common.

(b) Temperate deserts like Mojave in Southern California where day-time temperatures are very hot in summer but cool in winters.

(c) Cold deserts like the Gobi desert in China has cold winters and warm summers.

Desert plants and animals are having most typical adaptations for conservation of water. Many desert plants are found to have reduced to scaly leaves so as to cut down loss of water due to transpiration or have succulent leaves to store water. Many a time their stems get flattened and develop chlorophyll so that they can take up the function of photosynthesis. Some plants show very deep roots to tap the groundwater. Many plants have a waxy, thick cuticle over the leaf to reduce loss of water through transpiration. Desert animals like insects and reptiles have thick outer coverings to minimize loss of water. They usually live inside burrows where humidity is better and heat is less. Desert soil is rich in nutrients but deficient in water. Due to low species diversity, shortage of water and slow growth rate, the desert plant communities, if faced with a severe stress take a long time to recover.

Ecosystems in Aquatic Conditions

Aquatic ecosystems dealing with water bodies and the biotic communities present in them are either freshwater or marine. Freshwater ecosystems are further of standing type (lentic) like ponds and lakes or free-flowing type (lotic), like rivers. Let us consider some important aquatic ecosystems.

(a) *Pond Ecosystem:* It is a small freshwater aquatic ecosystem where water is stagnant. Ponds may be seasonal in nature i.e. receiving enough water during rainy season. Ponds are usually shallow water bodies which play a very important role in the villages where most of the activities centre around ponds. They contain several types of algae, aquatic plants, insects, fishes and birds. The ponds are, however, very often exposed to tremendous anthropogenic (human-generated) pressures. They are used for washing clothes, bathing, swimming, cattle bathing and drinking, etc. and therefore get polluted.

(b) *Lake Ecosystems:* Lakes are usually big freshwater bodies with standing water. They have a shallow water zone called Littoral zone, an open-water zone where effective penetration of solar light takes place called Limnetic zone and a deep bottom area where light penetration is negligible, known as profundal zone.

The Dal Lake in Srinagar (J & K), Naini Lake in Nainital (Uttarakhand) and Loktak lake in Manipur are some of the famous lakes of our country.

Organisms : The lakes have several types of organisms:

(a) Planktons that float on the surface of waters e.g. phytoplanktons like algae and zooplanktons like rotifers.

(b) Nektons that swim e.g. fishes.

(c) Neustons that rest or swim on the surface.

(d) Benthos that are attached to bottom sediments e.g. snails.

(e) Periphytons that are attached or clinging to other plants or any other surface e.g. crustaceans.

Stratification : The lakes show stratification or zonation based on temperature differences. During summer, the top waters become warmer than the bottom waters. Therefore, only the warm top layer circulates without mixing with the colder layer, thus forming a distinct zonation:

Epilimnion : Warm, lighter, circulating surface layer

Hypolimnion : Cold, viscous, non-circulating bottom layer.

In between the two layers is thermocline, the region of sharp drop in temperature.

Types of Lakes : Some important types of lakes are:

(a) Oligotrophic lakes which have low nutrient concentrations.

(b) Eutrophic lakes which are overnourished by nutrients like nitrogen and phosphorus, usually as a result of agricultural run-off or municipal sewage discharge. They are covered with "algal blooms" e.g. Dal Lake.

(c) Dystrophic lakes that have low pH, high humic acid content and brown waters e.g. bog lakes.

(d) Endemic lakes that are very ancient, deep and have endemic fauna which are restricted only to that lake e.g. the Lake Baikal in Russia; the deepest lake, which is now suffering a threat due to industrial pollution.

(e) Desert salt lakes that occur in arid regions and have developed high salt concentrations as a result of high evaporation e.g. great salt lake, Utah; Sambhar lake in Rajasthan.

(f) Volcanic lakes that receive water from magma after volcanic eruptions e.g. many lakes in Japan. They have highly restricted biota.

(g) Meromictic lakes that are rich in salts and are permanently stratified e.g. lake Nevada.

(h) Artificial lakes or impoundments that are created due to construction of dams e.g. Govindsagar lake at Bhakra-Nangal.

Ecosystem in Streams

These are freshwater aquatic ecosystems where water current is a major controlling factor, oxygen and nutrient in the water is more uniform and land-water exchange is more extensive. Although stream organisms have to face more extremes of temperature and action of currents as compared to pond or lake organisms, but they do not have to face oxygen deficiency under natural conditions.

This is because the streams are shallow, have a large surface exposed to air and constant motion which churns the water and provides abundant oxygen. Their dissolved oxygen level is higher than that of ponds even though the green plants are much less in number. The stream animals usually have a narrow range of tolerance to oxygen. That is the reason why they are very susceptible to any organic pollution which depletes dissolved oxygen in the water. Thus, streams are the worst victims of industrial development.

River Ecosystem: Rivers are large streams that flow downward from mountain highlands and flowing through the plains fall into the sea. So the river ecosystems show a series of different conditions. The mountain highland part has cold, clear waters rushing down as water falls with large amounts of dissolved oxygen. The plants are attached to rocks (periphytons) and fishes are cold-water, high oxygen requiring fish like trouts. In the second phase on the gentle slopes, the waters are warmer and support a luxuriant growth of plants and less-oxygen-requiring fishes.

In the third phase, the river waters are very rich in biotic diversity. Moving down the hills, rivers shape the land. They bring with them lots of silt rich in nutrients which is deposited in the plains and in the delta before reaching the ocean.

Ecosystem in Oceans

These are gigantic reservoirs of water covering more than 70% of our earth's surface and play a key role in the survival of about 2,50,000 marine species, serving as food for humans and other organisms, give a huge variety of sea-products and drugs. Oceans provide us iron, phosphorus, magnesium, oil, natural gas, sand and gravel. Oceans are the major sinks of carbon dioxide and play an important role in regulating many biogeochemical cycles and hydrological cycle, thereby regulating the earth's climate.

The oceans have two major life zones:

Coastal zone with relatively warm, nutrient rich shallow water. Due to high nutrients and ample sunlight this is the zone of high primary productivity.

Open Sea: It is the deeper part of the ocean, away from the continental shelf (The submerged part of the Continent). It is vertically divided into three regions:

(i) Euphotic zone which receives abundant light and shows high photosynthetic activity.

(ii) Bathyal zone receives dim light and is usually geologically active.

(iii) Abyssal zone is the dark zone, 2000 to 5000 metres deep. The abyssal zone has no primary source of energy i.e. solar energy. It is the world's largest ecological unit but it is an incomplete ecosystem.

Ecosystem in Estuary

An estuary is a partially enclosed coastal area at the mouth of a river where fresh water and salty sea water meet. These are the transition zones which are strongly affected by tidal action. Constant mixing of water stirs up the silt which makes the nutrients available for the primary producers.

There are wide variations in the stream flow and tidal currents at any given location diurnally, monthly and seasonally. Therefore, the organisms present in estuaries show a wide range of tolerance to temperature and salinity. Such organisms are known as eurythermal and euryhaline. Coastal bays, and tidal marshes are examples of estuaries.

Estuaries have a rich biodiversity and many of the species are endemic. There are many migratory species of fishes like eels and salmons in which half of the life is spent in fresh water and half in salty water.

For them estuaries are ideal places for resting during migration, where they also get abundant food. Estuaries are highly productive ecosystems. The river flow and tidal action provide energy subsidies for the estuary thereby enhancing its productivity. Estuaries are of much use to human beings due to their high food potential. However, these ecosystems need to be managed judiciously and protected from pollution.

Climatic Factors

Information about the prevailing climate is very important in deciding the ecosystem of a particular area. With the exception of those in the deep ocean, ecosystems are all at least partly shaped by climate.

Climate: Prevailing weather conditions at a location - e.g., temperature, humidity, wind speed and direction, cloud cover and rainfall.

Climate is the outcome of a number of factors:

1. Variations in the amount of incoming solar radiation.
2. The earth's daily rotation and its path around the sun.
3. Distribution of continents and oceans.
4. Elevation of land masses.

These factors interact to produce prevailing winds, ocean currents, and seasonal patterns of temperature and rainfall.

Global air currents influence patterns of temperature and rainfall by the following mechanism.

1. Sun's rays are more concentrated on the equator - as air near the equator heats up, it picks up moisture and begins to rise.
2. As the moist air rises, it cools and releases much of its moisture as rain - high rainfall in the tropics.
3. This drier air moves north and south away from the equator - begins to descend at about 30 degrees north and south latitude and is heated adiabatically as it descends by increasing atmospheric pressure - these descending masses of hot, dry air produce deserts at these lantudes.
4. Some of the air from these descending air masses is diverted back towards the equator, while part continues to move away from equator.
5. Air masses pick up moisture and are heated at 60 degrees latitude - rise and release moisture as rain - e.g., moist climate of Pacific northwest.

Some of this air moves south and some moves north, where it cools and descends on the polar regions - this is dry, cold air and polar regions are actually deserts.

The movement of these air masses away from the equator are also influenced by the earth's rotation and the formation of high and low pressure centers over continents and oceans - these factors interact to produce the directions of regional prevailing winds.

Prevailing winds, the earth's rotation and variations in ocean water temperature produce ocean currents which also have distinct regional patterns. The topography of continents (e.g., mountains, valleys, etc.) influence regional climatic patterns - a rain shadow effect.

Various types of ecosystems found in the world can also be classified. The most basic grouping is based on whether or not the ecosystem is on land (terrestrial) or in the water (aquatic).

Terrestrial ecosystems – regional ecosystem types having similar communities are sometimes called biomes. Biomes arise through a process called succession i.e. changes in community composition through time, Primary succession - complete community development from a bare rock surface- involves formation of soil from rock and may take thousands of years to complete - e.g., lava flow to tropical rainforest, a bare area generated after landslide.

Secondary succession - In it existing plant community is removed by a disturbance (e.g., fire) - doesn't include soil formation - time scale is 10-100 years.

In succession, the first community is called the pioneer community and it generally consists of lichens, mosses, weeds, etc. which are able to colonize disturbed areas - gradually each plant community modifies the environment, making it unsuitable for itself, but suitable for another community. In this process gradually each community is replaced by another-process culminates when they reach the climax community which is the final stage in succession.

Climax Communities: persistent, stable collections of species that typify particular regions of the world - climax communities are called biomes by ecologists.

Characteristics of biomes are determined by regional climate and by the types of soils available for plants to grow on.

Examples of some of the major biomes on earth are:

1. Tundra - cold temperatures for much of year - short-growing season (60 days)- may be alpine or arctic - low productivity because of short-growing system-decomposition slow because of cold temperatures - plant litter accumulates and soils have high organic content - low plant species diversity - only grasses, dwarf shrubs grow.

2. Forests - tremendous above-ground biomass - about 40 times more' as productive than tundra - competition for light important to plants - only 1% of sunlight hitting canopy reaches forest floor — forests can be subdivided into:
   A. Coniferous (evergreen) forests - cool, wet climates - pines, firs, spruces are common - found in northern latitudes.
   B. Temperate deciduous forest - cold regions - e.g., beech, maple, oak, hickory- deciduous trees lose and regrow leaves yearly.
   C. Tropical rainforest - most productive of all ecosystems - moderate temperatures and high rainfall (e.g., > 200 cm. per year) - decomposition rates are high - low organic matter in soil - soils have low fertility because nutrients are tied up in tremendous biomass of forest - highest plant and animals species diversity of any biome- e.g., a single acre may contain over 400 different species of trees - rainforests are being destroyed and an estimated 1 species/hour is going extinct due to the anthropogenic disturbances of rainforest.
3. Grasslands - 10"-30" rainfall/year - not enough rainfall to support trees -droughts are common - soils very fertile and high in organic matter - grasses are dominant plants and animal community dominated by grazers- prairie of North America- once supported 65 million bison and probably an equal number of pronghorn antelope.
4. Savannas— essentially grassland with scattered trees and shrubs - Africa, Australia and South America - animals consist of grazers (e.g., zebras, rhinos, antelopes) and browsers (elephants, giraffes) - trees deeply rooted and act as nutrient pumps - leaves fall off trees and decompose to make nutrients available to grasses -

trees are being used for fuel by humans which is drastically decreasing soil productivity.

5. Shrublands - dominated by shrubs 15'-20' tall - example is the chaparral of southern California - found around Mediterranean and west coast of South America - common feature is prolonged summer drought - shrubs are evergreen with thick, waxy leaves to prevent water loss - fire is an important part of this biome - it may have a 40 year cycle - fire suppression by human interference has resulted in more accumulation of biomass so the fires are often larger and more destructive when they occur.
6. Deserts - less than 10" rainfall/year - high temperatures increase evaporative water loss - least productive terrestrial ecosystem - sensitive to disturbance - plant reproduction may be a rare event - e.g., stand of mesquite in which all individuals are 250 years old - little or no reproduction in 250 years - desert plants and animals display numerous water conserving adaptations.

Aquatic Ecosystems: These can be classified into freshwater and marine ecosystems. Freshwater ecosystems include streams (lotic ecosystems), lakes and ponds (lentic ecosystems) and marshes and bogs (wetlands).

Lotic ecosystems - moving water - speed of water current important for different aquatic species because it determines rate at which materials such as nutrients are flushed out of the system, it also determines the type of bottom (e.g., sandy, gravel, or rock), which, in turn, determines the kinds of plants and animals that can live there.

Lotic systems are largely open systems as water is constantly moving - nutrients such as N and P are limiting because they are flushed out - major input of carbon is in the form of leaf litter from nearby terrestrial ecosystems. Lentic systems - lakes and ponds - standing water - more closed than lotic systems and have lower rates of nutrients loss.

Lakes and ponds can also be divided into three life zones

1. Littoral zone - closest to shore - contains rooted aquatic plants
2. Limnetic zone - sun lit body of lake
3. Profundal zone - below level of light penetration - no photosynthesis - decomposition is major process in this zone.

Lakes in temperate zones of the world display thermal stratification in the summer- surface layer (epilimnion) is heated by sun - temperature drops quickly in middle layer (thermocline) - bottom layer (hypolimnion) - these layers do not mix - epilimnion becomes nutrient poor as phytoplankton photosynthesize and use up nutrients - detritus (organic matter) sinks to bottom of lake - hypolimnion becomes oxygen-depleted as organic matter is decomposed.

Turnover occurs in fall as epilimnion cools, gets heavier and sinks—results in redistribution of oxygen and nutrients.

Lakes classified according to their nutrient status:

1. Oligotrophic - nutrient poor - low organic matter - low phytoplankton productivity.
2. Eutrophic - nutrient rich - high organic matter, high phytoplankton productivity- oligtrophic lakes can become eutrophic through inputs of nutrients from outside by human activity - e.g., sewage pollution, agricultural runoff.

Marine Ecosystems - begin at ocean's edge.

Estuaries - formed where large river flows into ocean - mixing zone of freshwater and saltwater - organisms must be adapted to changing salinity - they act as a nutrient trap - trap nutrients brought in by tides and carried down by river - because of their rich nutrients, estuaries have very high primary productivity - estuaries are often called "nurseries of the sea" because they are the place where many larval and immature marine fish grow up and develop before moving out into the ocean as adults. Seashores -exposed to waves and tides - classified as rocky or sandy.

Rocky shore - offer something solid for plants and animals to attach to - seaweed (algae) are primary producers - animals include barnacles, oysters, sea urchins. Sandy beach - most organisms dwell under the sand - e.g., ghost crabs, ghost shrimp, sandworms.

Coral reefs - diverse, productive ecosystems - found in shallow, tropical waters - reefs formed from calcium carbonate (limestone) exoskeletons of corals - crevices and caves provide shelter from many different plants and animals - sponges, echinoderms, fishes.

Ocean - covers roughly ¾ of earth's surface - can be divided into 3 life zones:

1. Neritic zone - shallow coastal waters - sun_2 light - seaweds and phytoplankton abundant.
2. Pelagic zone - open sea - includes sunlit waters - water is nutrient poor and productivity is comparable to that of the terrestrial deserts - phyto- and zooplankton base of trophic structure - fed upon by fishes and whales.
3. Benthic - ocean floor - temperature about 2 degree Centigrade and pressures are intense - total darkness - bombarded by dead, organic matter sinking from above- instead of ecosystems based on photosynthesizing autotrophs, benthic ecosystems are mostly decomposers and organisms that feed on decomposers (detritivores).

Benthic habitats are also the location of a unique, recently discovered ecosystem.

The earth's surface consists of large tectonic plates - in areas of the ocean bottom where these plates are formed, heated water is released from cracks in the earth's crust- these areas are called hydrothermal vents - ecosystem associated with these vents are based not on photosynthetic plants (no light), but on chemosynthetic bacteria that are able to obtain energy from hydrogen sulfide (H2S) released at the vents, these bacteria are the primary producers are fed upon by various consumers (worms and other invertebrates).

Gibbs Free Energy: The value of Gibbs Free Energy for a reaction is based upon its enthalpy and entropy. Both of these factors can have either a negative or positive effect on whether or not a reaction will occur spontaneously or at all. It can help to predict at what temperature a non-spontaneous reaction will become spontaneous.

It is called free energy because it is the energy that will be released or freed up to do work. Gibbs Free energy is defined as:

$$G^\circ = \Delta H^\circ - T\Delta S^\circ$$

It is also defined as G° = Gproducts - Greactants. But this equation is only valuable if the reaction is carried out at standard temperature and pressure. Any other temperature requires the use of the first equation. A change can only be spontaneous if it is accompanied by a decrease in free energy. In other words, for a change to be spontaneous, G must be negative. When a change is exothermic and is also accompanied by an increase in entropy, both factors favour spontaneity. i.e., H is negative (- ve) S is positive (+ ve) G = H - TS = (-) - T(+)

In such a change, G will be negative regardless of the value of the absolute temperature, T (which can only have positive values). Therefore, the change will occur spontaneously at all temperatures. On the other hand, if a change is endothermic and is accompanied by a decrease in entropy, both factors work against spontaneity. i.e., H is positive (+ve) S is negative (-ve) G = H - TS = (+) - T(-)

In this case, G will be positive at all temperatures and the change will always be non-spontaneous. When H and S have the same sign, the temperature becomes critical in determining whether or not an event is spontaneous.

If H and S are both positive, G = (+) - T(+)

Only at relatively high-temperatures will the value of TS be larger than the value of H so that their difference, G, is negative. A familiar example is the melting of ice.

$H_2O(s)$) $H_2O(1)$

Here is a change that we know is endothermic and occurs with an increase in entropy. At temperatures above 0°C (when the pressure is 1 atm), ice melts because the TS term is bigger than the H term. At lower temperatures, ice doesn't melt because the smaller value of T gives a smaller value for TS and the difference H-TS, is positive.

For similar reasons, when H and S are both negative, G will be negative only at relatively low temperatures. The freezing of water is an example.

$H_2O(1)$) $H_2O(s)$

Energy is released as the solid is formed and the entropy decreases. Water freezes spontaneously at low temperatures, that is, below 0°C.

Redox Potential

To understand it one should know the basics of redox potentiometry.

***Fundamental of Redox Potentiometry*:** Oxidation reduction (or redox) reactions involve the transfer of an electron (or H, or hydride group) from a donor to an acceptor. The donor loses an electron, and changes from the reduced to the oxidized form; the acceptor gains an electron, and changes from the oxidized to the reduced form. The oxidized and reduced form of a redox component are said to form a redox couple.

The special interest of redox reactions in thermodynamics lies in the fact that the free energy change can be measured directly using an electrochemical cell

Overall reaction:

$A^{-}+B \Leftrightarrow A+B$

In an electrochemical cell the components of the reaction

are separated so that the donor couple and acceptor couple are in separate reaction vessels. A salt bridge connects both vessels, so that the electrostatic potential difference between them is zero. In each vessel, an electrode (usually made of Pt) connects the reaction mixture in one cell to that in the other, through the measuring device (a potentiometer or electrometer).

The electrochemical cell therefore consists of two half-cells, each of which contains the reactants (oxidized and reduced components) of one redox couple, making up the half-cell reactants. In principle, the circuit allows the electrons to flow from one half-cell to the other, down the gradient provided by the difference in free energy of the electrons (measured in volts) in equilibrium with the two half-cell reactions. By channeling this flow through the measuring device, the system is constrained so that the flux is zero (or very nearly so).

If a potentiometer is used, the condition of zero flux is obtained by balancing the voltage difference between the two half-cell reactions through an applied potential of opposite sign, adjusting the current, measured with a sensitive galvanometer (or other suitable device), to zero. If an electrometer is used, the voltage difference between the half-cell reactions is measured directly, the condition of zero flux being achieved by use of an electrometer of very high input impedance.

In either case, the voltage measured shows the driving force represented by the difference in free energy of the electrons in equilibrium with the two half-cell reactions. This difference is called the electrode potential difference, the oxidation reduction potential difference, or the redox potential difference.

Because the redox potential difference is measured under conditions of zero flux (the "reversible" conditions of classical thermodynamics), it is a direct measure of the free energy change of the reaction of the electrochemical cell. The electrode potential measures the intensity factor (DE, measured in volts) of the work term (DG′ = -zFDE) for the process represented by the oxidation reduction reaction of the electrochemical cell.

In the case of the potentiometric measurement, this can be seen clearly because the voltage supplied by the potentiometer balances the electrode potential difference by application of external work from a change in state in the surroundings (the potentiometer).

The Reference Electrode : In practice, the redox potential difference is measured with respect to a standard electrode. Several sorts of electrodes are commonly used. The standard electrode is a hydrogen half-cell with the reaction:

$H_2(gas) \Leftrightarrow 2H^+ + 2e^-$

in which all components are in their standard states (1 atm pressure for the gas, 1M activity for the proton, or pH 0). This is not a convenient device to handle, so more often a secondary standard reference electrode is' used, a calomel electrode or a Ag/AgCl electrode. These latter have redox potentials with respect to the standard H_2-electrode which are well characterized. The redox potential for a couple is usually assumed to be the value with respect to the standard reference electrode (the hydrogen electrode). If a value is measured with respect to a secondary standard, the secondary reference electrode potential with respect to the standard is added to this value.

Free energy change for an oxidation reduction reaction We will consider a redox reaction

$A^- + B \Leftrightarrow A + B^-$

where the two half-cell reactions are those of the diagram above. We will assume that the B/B^- couple is a standard reference electrode.

Let us consider the half-cell reaction for the couple A/A,

$A^- \Leftrightarrow A + e^-$

and suppose that at activities [A] and [A^-], an electrode potential of E volts is measured with respect to the reference half-cell. If the concentrations are changed to [A] and [A^-], a different electrode potential of E′ volts is measured. Under the conditions of measurement (zero flux), the reaction of eq. 1 is at equilibrium, in each case, with the free energy of the electron measured by the electrode potential. Chemical free energy of A^- = chemical free energy of A + electrical free energy of e^-

or more concisely (and rearranging terms)

$$m_A - m_{A-} - zFE = 0 = m'A - m'A- - zFE'$$

Here z is the number of electrons involved in the reaction (in this case 1), and F is the Faraday constant. (The sign of E is negative because the electron carries a negative charge.)

Relation between Electrode Potential and Free Energy of Reaction : The complete reaction of an electrochemical cell can be treated like any other reaction, using the equation for DG'. Since the activity terms in the logarithmic ratio are the same as the activity terms in the equations for E above, it is a straightforward exercise to substitute between equations to find the relation between E and DG. This gives the following relationships:

$$DG' = - zFDE'$$

$$DG^{\circ} = - zFDE^{\circ}$$

$$DG^{\circ\prime} = - zFDE^{\circ\prime}$$

Analysis on Gravimetric

Gravimetric analysis a method of quantitative chemical analysis in which the constituent sought is converted into a substance (of known composition) that can be separated from the sample and weighed.

The steps commonly followed in gravimetric analysis are (1) preparation of a solution containing a known weight of the sample, (2) separation of the desired constituent, (3) weighing the isolated constituent, and (4) computation of the amount of the particular constituent in the sample from the observed weight of the isolated substance.

Of the many methods that may be employed for isolating the desired constituent from a solution of a sample, the most common is precipitation—that is, transformation into a substance not soluble in the solution. A reagent is added that forms an insoluble compound with the desired constituent but will not precipitate other constituents of the sample. The precipitate obtained is separated by filtration, washed free of soluble impurities, dried or ignited to remove water, and weighed.

Certain substances can be separated by virtue of their easy convertibility into gaseous compounds, as in the determination of carbonate in a mineral analysis. The sample is treated with an acid, and carbon dioxide is evolved as a gas. The gas is absorbed on a weighed quantity of a solid alkaline reagent, and the amount of carbon dioxide is determined from the gain in weight of the absorbent.

Electrodeposition is used in order to separate certain metals that can be plated out by passing an electric current through a solution of their salts. Copper in alloys may be determined by this method as long as the sample is free from other metals that plate out under the same conditions.

Errors made in gravimetric analyses usually relate to the purity of the isolated constituent. In general, the compounds that are precipitated are very insoluble, and negligible error results from the incompleteness of precipitation. Obtaining a precipitate that is 100 per cent pure and is exactly of the composition represented by a chemical formula is, however, considerably more difficult. All gravimetric methods are subject to some degree of error because of this difficulty.

Titrimetric Analysis

Types of Titrimetric Analysis

a. Acid-base Titrations

b. Redox Titrations

Titrations of Acid Base

Acid can be defined as a reagent which donates protons (H+) and a base as one which accepts protons. An acid-base reaction therefore involves the transfer of one or more protons. Consider the following reaction:

$HCl + NaOH = NaCl + H_2O$

The important thing here is that one "molecule" of HCl reacts with one molecule of NaOH. (Actually both are ionised in solution but for our purposes the above description is adequate). In terms of a titration, we might have the HCl in the conical flask (the titrand) and NaOH in the burette (the titrant).

Example: Suppose that the NaOH was in the burette and that this had been earlier standardised (i.e. its exact concentration is known) and is 0.0985 mol dm^{-3}. There are 25.0 cm^3 of HCI in the conical flask (delivered by pipette). The equivalence point was reached when 27.25 cm^{-3} of NaOH was added, what is the concentration of HCI ?

First Stage: How many moles of NaOH were required to reach equivalence point?

The concentration of the NaOH is 0.0985 mole dm-3 and 27.25 cm^{-3} were added so the number of moles added was:

$= (0.0985 \times 27.25)/1000 = 2.684 - 10^{-3}$

Since one molecule NaOH reacts with one molecule HCI (likewise one mole and one mole), there must be the same number of moles of HCl as there are NaOH = 2.684×10^{-3}.

OK, this is the number of moles of HCI is the conical flask i.e. in 25.0 cm^{-3}. Usually you are asked for concentration in terms of mole dm ^{3} so this figure must be multiplied by 1000/25. (Concentration volume required/concentration volume actual). So the concentration ($moldm^{-3}$) is ?

Exercises

1. A solution of HCI (25.0 cm^3) was titrated against NaOH (0.102 mole dm^{-3}) given the titre was 25.40 cm^3, calculate the concentration of the HCl solution.

   (Answer: 0.104 $moldm^{-3}$)

2. In another titration using the NaOH from above (0.102 mole dm^{-3}) against sulphuric acid (25.0 cm^3) the titre was 28.60 cm^3, calculate the concentration of the sulphuric acid.

   (Answer: 0.058 mole dm^{-3})

3. Arsenic (V) acid (H_3AsO_4), is a tribasic acid, 25.0 cm^3 of a solution of this acid required 35.7 cm^3 of sodium hydroxide (0.100 $moldm^{-3}$) for titration. Calculate the concentration of H_3AsO_4. (Answer: 0.0476 $moldm^{-3}$ or 4.76×10^{-2} $moldm^{-3}$).

4. If a monoacidic base of concentration 0.1000 mol dm^{-3} is titrated against 25 cm3 of a dibasic acid (0.0460 moldm-3) what volume of base will be required to reach equivalence point ? (Answer: 23.0 cm^3)

Redox Titrations

"Redox" stands for reduction - oxidation. In this type of reaction one species is reduced (that is a reduction in its oxidation number - or a gain of electrons for example M^3+ going to M^2+ whilst the other is oxidised (an increase in its oxidation number or a loss of electrons for example M^2+ going to M^3+).

Oxidation number is defined as the charge which an atom appears to have (assuming the bonding to be entirely ionic) when electrons are counted using the following rules.

Electrons shared between like atoms are equally divided between them. Hence all elements e.g., H_2, O_2 etc. have an oxidation number (or state) of zero.

Electrons shared between unlike atoms are counted with the more electronegative atom e.g., in HCl the electrons are assumed to be associated with the chlorine so the oxidation numbers are hydrogen +1, chlorine -1.

To make it easier use the following rules; The alkali metals (Group I) always are +1 Group II metals are +2.

Oxygen is -2 (except in peroxides)

Hydrogen is +1 (except with Group I and Group II when it is -1)

Example: ON of Mn in $KMnO_4$.

K = +1; 4 × O = (4 x -2) = - 8 So for electrical neutrality the ON on Mn must be +7.

Balancing Redox Equations

Oxidation involves an increase in ON and a loss of electron(s)

Reduction involves a decrease in ON and a gain of electron(s)

Oxidising agents cause a decrease in ON and the agent gains e^-'s. Reducing Agents cause an increase in ON and the agent loses e^-'s. There are a number of methods used in balancing redox reactions.

Oxidation Number Method

1. Assign ON's to substances for which ON changes.
2. Decide on the number of electrons to be moved per atom.
3. Decide on the number of electrons to be moved per formula unit.
4. Since electrons gained by the oxidisable reactant must be supplied by the reducible reactant, compensate electron gain or loss by inserting appropriate coefficients to the oxidising and reducing agents. In other words, get the ratio of the reacting species correct in terms of the electrons gained or lost.
5. Insert other coefficients to be consistent with the conservation of matter. Let's take a look at an example:

The reaction between $K_2Cr_2O_7$ and H_2SO_3 in acid solution to give $KHSO_4$ The basic "equation" might be written as:

$K_2Cr_2O_7 + H_2SO_3 + H+ = KHSO_4 + Cr^{3}+ H_2O$

For Cr:

1. For the reactant ($K_2Cr_2O_7$) the oxidation number on Cr = +6 For the product (Cr^{3+}; the oxidation number on Cr = +3
2. So we need 3 electrons (+6 goes to +3) for each Cr atom.
3. Since there are two Cr per formula unit we need 6 electrons per formula unit.

For S:

1. In H_2SO_3 ON for S = +4
   In $KHSO_3$ ON for S = +6
2. So for each S atom undergoing change there are 2 electrons released therefore 2 electrons per formula unit.
3. We now know that $K_2Cr_2O_7$ and H_2SO_3 must react in the ratio 2:6 or 1:3 We can now write;

$$K_2Cr_2O_7 + 3\,H_2SO_3 + H^+ = 3\,HSO_4^- + 2\,Cr^{3+} + 2K^+ + H_2O$$

We now need to balance H and O.

On the left hand side we have 16 O and on the right we have 12 O, so we need $4H_2O$ to make up the difference. This leaves an imbalance of H+ which can be made up simply be adding H+ to the lefthand side.

$$K_2Cr_2O_7 + 3\,H_2SO_3 + 5\,H^+ = 3\,HSO_4 + 2\,Cr^{3+} + 2K^+ + 4\,H_2O$$

Examples of some important Redox Titations:

Peroxydisulphate Ion

The peroxydisulphate ion (or persulphate ion) $S_2O_8^{-2}$ (usually the potassium or ammonium salt) is a strong oxidising agent in acid medium:

$$S_2O_8^{2-} + 2e^- > 2SO_4^{2-} \quad E = +2.0V$$

Any excess of this oxidant may be removed by boiling;

$$5_2O_8^{2-} + 2H_2O > SO_4^{2-} + 2H^+ + H2O2 \quad 2H_2O_2 > H_2O + O2$$

The important uses of this oxidant include the conversion of manganese to permanganate, chromium to dichromate and vanadium to vanadate.

Perchloric Acid

Hot concentrated perchloric acid is a strong oxidant;

$2ClO_4^- + 16H+ +14e^- > C1_2 + SH_2O$ E = +2V

Used to oxidise chromium and vanadium ores/alloys to dichromate and vanadate. Permanganate as an oxidising agent.

The permanganate (or manganate (VII)) ion is one of the most common oxidising titrant. In strongly acidic media, permanganate undergoes a five electron reduction.

$MnO_4^- + 8H^+ + 5e^- > Mn^{2+} + 4H_2O$ E = +1.5V

In weak acid, neutral or alkaline conditions permanganate undergoes a three electron transition to manganese (IV)

$MnO_4^- + 4H^+ + 3e^- > MnO_2 + 2H_2O$ E = +1.69V

Or in alkaline solution;

$MnO_4 + 2H_2O + 3e^- > MnO_2 + 4OH^-$ E = + 0.57V

Note: permanganate solutions are unstable, therefore not a primary standard.

$4MnO_4^- + 2H_2O > 4MnO_2 + 4OH^- + 3O_2$ *(MnO_2 catalyses this reaction)* Permanganate must be standardised for example with oxalate;

$H_2C_2O_4 > 2H^+ + CO_2 + 2e$

Overall: $2MnO_4 + 5H_2C_2O_4 + 16H^+ > 2Mn^{2+} + 10CO_2 + 8H_2O$

Initially the reaction is slow but is catalysed by Mn 2+ so becomes more rapid. Permanganate titrations:

Hydrogen peroxide in acidic conditions;

$5H_2O_2 + 2MnO_4 + 6H^+ > 2Mn^{2+} + 5O_2 + 8H_2O$

Iron (II) in acidic conditions is easily oxidised by permanganate

$MnO_4^- + 5Fe^{2+} + 8H^+ > Mn^{2+} + 5Fe^{3+} + 4H^2O$

Dichromate Reactions : Dichromate ion is an oxidising agent:

$Cr_2O_7^{2-} + 14H^+ + 6e > 2Cr^{3+} + 7H_2O$ E = +1.33V

It can be prepared as a standard solution and so avoids the need to standardise as is the case with permanganate.

Iodine Methods

$I_2 + 2e^- > 21^-$ E = +0.54V

Value for E is intermediate can therefore be reduced or oxidised. Iodine can be reduced to iodide by for example As (III), Sn (II) whilst iodide can be oxidised to iodine by for example permanganate.

Thiosulphate:

In acid, neutral or slightly alkaline solutions;

$25_2O_3^{2-} > S_4O_6^{2} + 2e^-$

$2S_2O_3^{\ 2-} + I_2 > S_4O_6^{2-} + 2I^-$

strongly alkaline solutions;

$2S_2O_3^{2-} + 4I_2 + 10OH^- > 2S_4O_6^{2-} + 8I^- + 5H_2O$

Iodate

$IO_3^- + 5I^- + 6H^+ > 3I_2 + 3H_2O$

Hydrogen Peroxide : Strong oxidant even in alkaline conditions.

$H_2O_2 + 2H^+ + 2e^- > 2H_2O$ E = +1.77V

Danger of Chemical Carcinogenesis : In 1775, an English physician named Percival Pott observed that the high incidence of cancer of the scrotum in chimney sweepers was due to their exposure to coal tars.

A century later, scientists in Germany found a high incidence of bladder cancer among workers exposed to aromatic amines (chemicals used in the dyeing and pigment industry). Since then, researchers have identified many more substances that are known or suspected to cause cancer. These substances are referred to as carcinogens. Presently in 25-30,000 chemicals are manufactured at commercial level or generated as waste and emissions released during industrial processes. Now pollution of human environment is suspected as a major cause of cancer.

Factors Carrying Risk of Cancer : Cancer is often perceived as a random, and unpredictable disease. In fact, it's estimated that about 80 per cent of all cancers are caused by "behavioral" factors and are therefore under human control. Two of the leading risk factors associated with cancer are tobacco smoke and diet. Together they are responsible for about two thirds of all new cancer cases.

Tobacco smoke is the major cause of 80 to 90 per cent of lung cancer cases. It is also an important cause of bladder and breast cancer. Diet associated cancers include colon, rectum, breast and stomach cancer. For example, high fat consumption is considered a risk factor for colon and possibly breast cancer. Other well-established risk factors include work-related exposure, alcohol consumption, radiation including sunlight, drugs and viruses. By comparison, the role of chemical pollutants is unclear. Scientists have estimated that chemicals in the environment account for about two per cent of all cancer deaths. There is little direct evidence linking exposure to specific pollutants in the general environment and the incidence of cancer.

Determining the Cancer Risk or Carcinogenicity : To determine the potential of chemical pollutants to cause cancer, scientists rely primarily on laboratory studies in which animals are exposed to chemicals, and when available, population studies involving human exposure to chemicals.

Chemicals can be grouped into six categories: "carcinogenic to humans", "probably carcinogenic to humans", "possibly carcinogenic to humans", "unlikely to be carcinogenic to humans", "probably not carcinogenic to humans" and "unclassifiable" with respect to cancer causing ability. Scientists assess carcinogenicity by weighing the available data from both animal and human studies.

So far, out of thousands of chemicals tested, scientists have identified only a few hundred that cause cancer in rodents and around 50 that definitely cause cancer in humans.

The number of confirmed human carcinogens is lower because, for the majority of chemicals, there is often little or no information available on cancer incidence rates in humans. Moreover, effects that occur in laboratory animals will not necessarily occur in humans. In other words, what is harmful to a rat or mouse may not be harmful to a human.

Carcinogens in the Atmosphere

In theory, there is no "safe" level of exposure to a carcinogen. It is generally assumed, but has never been proven, that there is some chance of developing cancer at any level of exposure. In other words, contact between a single human cell and a carcinogen molecule may be enough to cause cancer. This does not mean that cancer is the inevitable or even the likely outcome of such an event.

Human cells have developed a range of defenses to guard against all sorts of injuries, including those that lead to cancer. As a result, cancer usually occurs only after prolonged and continuous exposure to carcinogens. In addition, as the exposure levels decrease, so does the risk of developing cancer.

People today are exposed to trace amounts of various chemicals in their air, water and food. But the levels are generally far below those levels associated with harmful effects. In the workplace, however, exposure to much higher levels of chemicals can occur.

In 1989, the government of Canada listed a group of 44 commercial chemicals, known as priority substances list, of the 44 substances on the List, 19 were considered "toxic" to human health. These substances were considered toxic largely based on their potential to cause cancer. The actual risk of a chemical causing cancer depends on two factors: its ability to cause cancer, and the levels to which people are exposed.

A partial list of substances known to be toxic or carcinogenic to human health is given below:

- Arsenic (Inorganic Arsenic)
- Benzene
- Benzidine
- Bis (2-Chloromethyl) Ether
- Cadmium (Inorganic Cadmium)
- Chlorinated Paraffin Waxes
- Chloromethyl Methyl Ether
- Chromium (Hexavalent Chromium)
- 3,3-Dichlorobenzidine
- 1,2-Dichloroethane
- Bis (2-Ethylhexyl) Phthalate (DEHP)
- Dichloromethane
- Hexachlorobenzene
- Mineral Fibres (Refractory Ceramic Fibre)
- Nickel ("oxidic", "sulphid", and "soluble" forms)
- Polycyclic Aromatic Hydrocarbons
- Polychlorinated Dibenzodioxins
- Polychlorinated Dibenzofurans
- Trichloroethylene

2

Endocrine Disruptor Chemicals

Fish Reproductive Health

There is growing scientific consensus that numerous industrial and agricultural chemicals have the ability to interfere with endocrine systems and hormonal activities of all animals including fish. These endocrine disrupting chemicals (EDCs) are thought to be especially important at the larval or developmental stages of fish, disrupting sexual development, behaviour and fertility.

EDCs are typically identified as compounds that can interact with oestrogen or androgen receptors and thus act as agonists or antagonists of endogenous hormones. Growing evidence shows that they may also modulate the activity/expression of steroidogenic enzymes. These are expressed not only in the gonads but also in many tissues that have the ability to convert circulating precursors into active hormones. In this way, EDCs may impact both on sexual differentiation and development of all animals.

NDTV, NewDelhi in July 2003 had brought out the information from Department of Environmental Biology of the Patna University that the toxic compounds have polluted the Ganga water and have contaminated the fish in river. Right from Haridwar to Faraka, the fish contain high amounts of pesticide and they pose a serious health hazards to those who eat the fish. Pesticide used in agriculture like DDT are non-biodegradable and their residual accumulation is high in Ganga water and Ganga water contains 0.38 ppb.

The *Labeo rohita,* one the most eaten fish in Bihar contains 182.3 ppb where as permissible limit for such toxic compounds for freshwater aquatic life is 0.0014 parts per billion (ppb). They contain too much quantity of organochlorine compound which may cause either cancer, impair the reproductive system and definitely affect the immune system leading to mortality.

The scientist in Centre of Marine Biotechnology Maryland found that compounds in a wide range of detergents, plastics, pesticides, medicines and even thalates disrupted the sexual development and juvenile zebrafish, *Danio rerio*. The scientists discovered that many more classes of environmental chemicals than suspected are functioning as EDCs, and these chemicals are interacting directly with genes that are critical for reproductive success.

Unlike most of animals, many fish produce two forms of a gene responsible for the enzyme aromatase, which in turns makes estrogen (Sundaray. J.K 2004). One form of aromatase is in the ovaries; the other is in the brain. The study result suggest that direct chemical disruption of the brain aromatase gene, which directly affects production of brain estrogen, may be key mechanism behind the endocrine-related disturbance of normal reproduction and development in fish.

Evidence is now accumulating that even low levels of pollutants can disrupt the functioning of the endocrine system of fish and fish can provide an early warning of effects that may later become apparent in other wildlife species and ultimately humans. Some of the problems found in fish including decreased fecundity, genital abnormalities, altered behaviour patterns and weak response to stress and disease are now observed in human populations.

Some argue that the damage to the environment is already widespread, while others point out that EDCs are far less potent than the bodies' own hormones, and in certain cases (e.g., environmental anti-estrogens and estrogens) could balance each other out. Despite this debate, there is little disagreement over the need to better understand environmental impacts of the above disrupting chemicals world wide. However in India there is no much information on the endocrine disruptor chemicals and its impact on fish maturation and reproduction particularly the effect at molecular level. This review will deal with the research work have been completed in India and abroad and will give an idea about future need of the research in India.

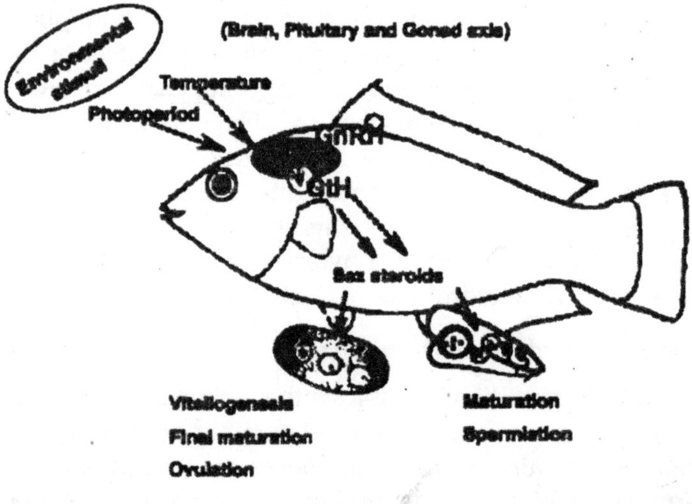

Fig.: Schematic representation of hormonal control of reproduction in fishes

Endocrine System and Steroidogenesis: Endocrine system is a interplay between number of glands and hormones that regulates many of the body's functions; including growth, development and maturation, as well as the way various organs operate. It is a control system of the body includes the endocrine glands the pituitary, thyroid, adrenal, thymus, pancreas, and hormone secreting organ like ovaries, and testes. The endocrine system release carefully-measured amounts of hormones into the bloodstream that act as natural chemical messengers, travelling to different parts of the body in order to regulate sexual development and the seasonal reproductive cycles and to evoke a stress response to external threat. At its core are the hypothalamus and pituitary, which respond to neural signals from the brain and convert them into hormone messenger which act on the individual glands. Once they have elicited the appropriate action in their target tissues, hormones are converted by the liver to metabolites which are more easily excreted. There is a negative feedback of the circulating hormones to the pituitary and hypothalamus which maintains the normal hormonal equilibrium.

The synthesis of sex steroids from cholesterol requires trafficking between mitochondria and smooth endoplasmic reticulum and involves many enzymatic steps. Most of these steps use cytochrome P450 haem-containing enzymes, and the genes coding for these enzymes are abbreviated to CYP. The enzyme that has received the most attention with regard to enzyme modulation and endocrine disruptors has been aromatase (CYP19) that converts androgens to oestrogens. Other enzymes have included 3 a-hydroxysteroid dehydrogenase (HSDs) and 17 a-hydroxysteroid dehydrogenase.

Concept of Endocrine Disruption

An endocrine disruptor is a synthetic chemical compounds when absorbed in the body that either mimic or antagonise the action of naturally occurring hormones are termed endocrine disrupters or endocrine disrupting compounds (EDCs). EDCs are defined as "an exogenous substance or a mixture, that alters function(s) of the endocrine system and consequently causes adverse health effects in a intact organism, or its progeny, or (sub) populations" (European Commission, 1996). Concern over the possible consequence of exposure to xenobiotic compounds able to modulate the endocrine system (hormones and hormone synthesising steroidogenic enzymes) has drawn the attention of International agencies, like European Commission, the European parliament, the US Environmental Protection Agency, the Organisation for Economic Cooperation and development, the WHO International programme on Chemical Safety, non-government organisation and the chemical industry (OECD, 1989, CSTEE,1999).

Mechanism of Endocrine Disruption

Exposure to endocrine disruptors can occur through direct contact with pesticides and other chemicals or through ingestion of contaminated water, food, or air. Chemicals suspected of acting as endocrine disruptors are found in insecticides, herbicides, fumigants and fungicides that are used in agriculture as well as in the home. Endocrine disruptors enter the air or water as a by-product of many chemical and manufacturing processes and when plastics and other materials are burned.

The biological actions of hormone synthesised within an organism, such as estrogen, progesterone, testosterone and thyroxine, are mediated by high affinity receptor proteins located within target cells. The interaction of a hormone with its receptor initiates a cascade of events that lead to the myriad of effects associated with particular hormone. Some exogenous chemicals may bind to a receptor and mimic or block the actions of its natural hormone. Such compounds include some naturally occurring chemicals, such as coumestroland genistein, pharmaceuticals, such as diethylstibestrol, 17a-ethinyloestradiol and tamoxifen and industrial chemical such as DDT, bisphenol-A and nonylphenol (DFG and Eisenbrand, 1996). Compounds that have been shown to alter estrogen biosynthesis are cyanoketone, ketoconazole and the fungicide fenarimol (Hirsch *et al.*, 1987). Methoxychlor, chlordecone (kepone), DDT, some PCBs and alkylphenols can disrupt estrogen receptor function (Mueller and Kim, 1978; White *et al.*, 1994). They can also affect the action of key enzymes of biosynthesis and metabolism that will alter circulating hormone levels or cause structural changes which affect the function of an endocrine gland.

In mammals, the steroidal sex hormones and their receptors have a regulatory role in developmental processes such as sex determination and differentiation. The development of the conceptus into a male phenotype is regulated by testicular hormones including testosterone.

The androgen receptor antagonists p,p0-DDE and vinclozin may demasculinise male rat pups following exposure of the pregnant dam (Kelce *et al.*, 1997). In birds, differentiation into a female phenotype is regulated by estrogen. Injection of an estrogen synthesis (aromatase) inhibitor into the fertilized egg may result in phenotypically sex-reversed females with bilateral testicles, sperm production and male copulatory behaviour (Elbrecht and Smith, 1992). In lower vertebrates and non-vertebrates, the mechanisms of sexual differentiation are less well characterised but may be more flexible than in mammals.

They are influenced by environmental factors such as temperature and xenogenous substances with hormonal activity (Campbell and Hutchinson, 1998). Experimentally disturbed sexual differentiation has also been reported in reptiles, amphibia, fish and non-vertebrates such as gastropods.

Fish as a Monitor

Endocrine disruption in fish appears more frequently associated with four types of land use: sewage treatment plants, pulp and paper mills, and areas of high industrial activity/ chemical contamination and house hold pollutants. The viability of commercial and ecologically important marine and freshwater fish stocks may be threatened by the feminisation of male fish by a range of EDCs. Scientists believe that there is already sufficient evidence to suggest that a precautionary approach should be adopted to try to minimise the effects of these chemicals now. As the aquatic ecosystem is a major recipient of pollutants, the fish constitute an important potential target and can be used as a research and bio-monitoring tool.

Fish have many advantages as experimental models in the study of endocrine disruption, and the underlying the basis of endocrine system is very similar to that of mammals. Fishes are ideal animal for assessing whole organism responses including the life cycles toxicity test as well as the fish early life stage toxicity test. The fish early life stage toxicity test leaves room for multigenerational studies as larva fish can be grown to maturity with subsequent monitoring of the reproductive success of second generation.

Recent evidence from the University of Newcastle suggests that there are estrogenic effects on wild population of marine fish exposed to sewage effluent. In marine waters, male flounder in the Tyne, Mersey and Solway estuaries have been found to show signs of feminisation. This has raised fears over the long-term effects on fish populations in waters receiving significant volumes of effluents. The research has suggested that "the incidence of the feminisation of male fish is higher than previously thought and is associated with discharges from sewage treatment works". Effects found have included:

- Egg-producing cells in the male testis
- Reduced testis growth rates and size
- Female reproductive tracts
- Increased liver size
- Increased levels of vitellogenin.

There is an increased concern to know about the EDCs and its impact on aquatic animal and environment. To cater this existing and evolving problems in aquatic environment and for the whole animal kingdom fish are good study subjects because:

- Their endocrine system is similar to those of higher vertebrates;
- Their susceptibility to EDCs has been validated by laboratory experiments;
- They are easy to work with; and
- There have been indications of endocrine disruption in fish in some areas.

Endocrine Disruptor Chemicals

Studies in the laboratory have shown that a number of synthetic and natural chemicals can interfere with the endocrine system in fish. These include certain chemicals used in industrial and manufacturing processes, classic contaminants such as polycyclic aromatic hydrocarbons (PAHs), polychlorinated biphenyls (PCBs), dioxins, certain pesticides, and a number of trace elements. The use of some fungicides (e.g., vinclozoilin and linuron) and herbicides also represents avenues for environmental exposure.

The human diet is also a potential source of EDCs especially for those who consume product rich in phytoestrogens. There are also emerging studies on the possibility that the use of Pharmaceuticals agents by live stock industry may be contributing to environmental loads of EDCs (e.g., Wilson *et al.*, 2002). An extensive chemical which are thought to be capable of disrupting the reproductive endocrine system of animals has been assembled.

Organochlorine Pesticides

Aldrin, chlorodecone (Kepone), 2, 4-D, DDE, DDT, dieldrin, endosulfan, endrin, a-and a -HCH (lindane), methoxychlor, mirex.

Oragnophosphate Pesticides

Chlorofenvinphos, cythion, diazinon, elsan, fenitrothion, fenthion, malathion, methyl parathion, mevinphos, monocrotophos, parathion, quinalhos, temephos, TEPA, tetrachlorviphos.

Industrial Chemicals

Alkypphenols, ammonia asbestos, chlorinated parafins, 4-chloroaniline, cyanide detergents, di-n-butyl phthalate, polyaromatiic hydrocarbon (PAHs; e.g. anthracene, benzopyrene, methylcholantherene, a-naphthoflavone), nitrate, nitrite, petroleum oil, phenol, pentachlorophenol, 4-nitrophenol, dinitro-o-cresol, polychlorinated biphenyls (PCBs; especially coplanar), polychlorinated dioxins, polybrominated naphthalenes, a-sitosterol, sulphide, thiourea, urea, acid water coal dust.

Herbicides

- Diquat dibromide, Atrazine, simazine, Diuron
- Plasticizers /surfactant
- Bisphenol A, dioxins, nonylphenol, Methoxychlor
- Fungicides
- Vinclozolin, prochloraz, imizalil, bifonazole, miconazole, clotrimazol
- Heavy metals
- Arsenic, cadmium, chromium, copper, iron, lead, mercury, tin, zinc
- Environmentally Induced abnormalities of EDCs on Fish.

The aquatic environment is subject to an ever increasing range of man made (anthropogenic or xenobiotic) pollutnats, reflecting the ever more rapid innovations of our technology to manufacture goods to satisfy a perceived increase in consumer demand on which our economy is based. Some of the pollutants that are now present in the tissue of fish, wild life and humans also reflect past usage of chemicals, such as organochlorine insecticides and PCBs. These chemicals have huge adverse impact on the reproductive health of fish, wild life and human as a whole.

Brain, Pituitary and Aromatase (Cyp 19) Gene Abnormalities

The function of the nervous system is the rapid communication of signals between different parts of the body, and the processing of these signals. The most complex part of the nervous system is the brain, where most of the processing occurs. Although most of the nervous system is protected from potential toxicants by the blood-brain barrier, some chemicals can penetrate this barrier (Anthony and Graham, 1991). A US EPA workshop concluded that a variety of endocrine disrupting chemicals, including some PCBs, dioxins, DDT and chlorinated pesticides, are capably of producing neurotoxicity (Kavlock *et al.*, 1996). Effects reported in the literature included changes in behaviour, learning and memory attention. 'Our Stolen Future' (Colborn *et al.*, 1996) describes several of these studies. Most of these toxic effects are believed to occur due to the EDCs influencing the development of the nervous system.

There two distinct type of aromatase are present in the teleost like P450arom-A (ovarian type) and P450arom-B (Brain type) (Sundaray *et al.*, 2006). Brain type aromatase have been targeted by some EDC or substances. It has been observed that nonylphenol an active endocrine disruptor substance targeting the expression pattern of estrogen receptors and aromatase is form genes in the brain.

Nonylphenol caused a significant induction of P450 arom and estrogen receptor isoform patterns in the brain and represent quantitative gene expression biomarker of exposure to endocrine disrupting chemicals in the environment. An exposure of 15 µg nonylphenol/L the brain P450aromB and P450aromA were induced 3.5- fold and 3.2-fold respectively in *Salmo salar* the Atlantic salmon. Similar induction response were also observed with ERá receptor (2.3-fold) and ERa receptor (2.3-fold) mRNA levels in the brain after 7 days exposure to 15 µg nonylphenol/L (Meucci V and Arukwe A. 2006). Organochlorine pesticides, such as lindane impurity, a-HCH mimic natural estrogen and induce a negative feedback response in the pituitary (Wester P.W *et al.*, 1985)

Y. Kazeto *et al.*, (2004) suggested that multiple classes of EDC have influences on the expression of two type of aromatase genes present in the zebrafish, *Danio rerio*, The expression of P450arom-ovarian type was significantly suppressed by ethylestradiol at high concentration but all other EDC (Nonylphenol, atrazine, clofibrate) treatment failed to modulate P450arom-ovarian type expression in zebrafish. Transcription of P450arom-brain type is strongly enhanced in zebrafish juveniles by a 3-day treatment with ethylestradiol and nonylphenol.

P450arom-brain type was significantly over expressed (3fold) in the brain of a fish *Rivulus marmoratus*, to EDC compounds like 4-NP, 4-tert-OP and bisphenol-A while compare to control (Y M Lee *et al.*, 2006). The expression of P450arom-ovarian type mRNA in gonads was also elevated following estrogenic exposure, which led to significant down regulation of P450arom-ovary type. Depressed P450arom-ovarain type activity was observed in the red mullet, *Mullus barbatus* when fishes were analysed from alkylphenol contaminant water during spring season (R. M. Skilton *et al.*, 2006).

Immune System

The immune system is responsible for resisting infectious agents, environmental substances and foreign or damaged cells (Dean and Murray, 1991). The capacity for immune response is determined early in development, so damage early in life will persist (Colborn, 1996).

The immune and endocrine systems are linked by a variety of signaling systems, and much research on wild animals has demonstrated that endocrine disrupting chemicals can reduce the effectiveness of the immune system (Kavlock *et al.*, 1996). A lot of this research has focused on marine animals and birds; for example, high levels of PCBs and DDT in the blood of dolphins was associated with decreased immune function and increased incidence of infection (Kavlock *et al.*, 1996). However less information is available in the EDC and its impact on immune system of fresh and marine water fish.

Female Reproductive Problem

Xenobiotic induced disruption of female fertility follows essentially the same pattern as that of male and can be caused by changes in brain-pituitary function, primary disruption of gamete structure and hormone secretion.

Ovarian Structure and Hormone

Abnormal development of the ovary can be caused by lack of stimulation by pituitary hormones, failure of steroid synthesis or direct cellular damage. The most commonly observed effect in the ovary is decrease in the number of large yolk eggs together with increased numbers of immature oocytes, suggesting a primary effect on pituitary function. Such changes have been observed with organochlorine (DDT, endosulfan, aldrin, chlordecone, a-BHC, methoxychlor), organophosphate and crabamate pesticides.

Synthesis of the steroid hormones testosterone and 17,20âP is identical in both ovary and testis and dependent upon the functional integrity of both the steroid producing enzymes and receptors for pituitary gonadotropin. Inhibition of ovarian steroid synthesis has been demonstrated for the pesticides DDT, a-BHC, malathion, metacid-50 and carbaryl. Hormones levels were also depressed in plasma of fish exposed to effluent of pulp mills or taken from the Great lakes and others areas rich in PCBs and PAHs (Kime, D.E, 1999). Ovary has an active aromatase enzyme that converts androstenedione to estrone in fish (Sundaray, et al 2004a). Inhibition of this aromatase activity in female fish will decrease estrogen synthesis, and have consequential effects on the ability of its liver to synthesise yolk proteins, leading to retarded growth of the oocytes. Such aromatase inihibiting activity has been demonstrated for imidazole fungicides (Monod. G. *et al.*, 1993), PAHs (Afonso, L.O.B *et al.*, 1997), pulp mills effluent (M. E. McMaster *et al.*, 1995) and tributyltin (TBT) (Fent. K, 1996). These compounds may also affect steroid feedback to the pituitary in both sexes since this is dependent upon aromatase activity.

Ovulation and Spawning

Ovulation is stimulated by a range of external stimuli such as presence of male and his courtship behaviour, spawning substrate, as well as temperature or water flow. Absence of current stimuli is frequent cause of ovulatory failure in captive fish. The final maturation of the eggs just prior to ovulation and spawning is dependent on a very rapid change in the pattern of secretion of gonadotrophin and ovarian hormones. Any disruption to this pattern can have major consequence for the ability to the fish to spawn. While long term exposure to. such chemicals would lead to delayed ovarian development.

Late spawning has been observed in water polluted by PCBs, pulp mill effluents or acids. These are usually a result of decreased steroid production causing slower ovarian development than normal, but even if the full complement of eggs is eventually spawned this delay may have serious consequence for the numbers of offspring which survive to form a breeding population (Kime. D. E, 1999).

Egg Number and Viability

Decreased production of yolk protein resulting from inhibition of either ovarian or liver function presents the fish with a similar choice to that resulting from decreased food availability.

It can produce the same number of smaller eggs, or a smaller number of eggs of normal size (C.R. Tyler and J. P. Sumpter, 1996) but the mechanism buy which that choice is made is unclear. Since the decreased nutrient content of small eggs with result in smaller larvae with a decreased survival rate, the net result in both cases may be decreased survival of the progeny.

Additional Effects on Female Fish

Xenobiotics can also affect secondary sexual characteristics in species in which they are present. The most documented such case is in mosquitofish and least killfish in which the anal fin of the male is modified to form gonopodium that is used to transfer sperm to the female during internal fertilization. Such development has also been observed in female fish captured below pulp mills in Florida and has been taken to indicate the presence of androgenic chemicals in the effluents (S.A Bortone *et al.*, 1989).

Since pulp mill effluents contain aromatase inhibitors (M.E. McMaster *et al.*, 1995) it is possible that this masculanisation could have been due to the absence of estrogen during sexual differentiation (Piferrer. F. *et al.*, 1994) rather than to the presence of a true androgen.

Male Reproductive Problem in Fish

Much of the impetus for recent upsurge in activity on endocrine disruption was the finding of decreased sperm counts and an increased incidence of genital abnormalities in humans (R. M. Sharpe and N. E. Skakkebaek, 1993). While some of these findings have been disputed, the much more extensive evidence from studies with fish adds considerable weight to the argument that such changes are occurring and that they are caused by exposure to anthropogenic pollutants rather than changes in human life style. Endocrine dysfunction in the testis may be apparent by changes in its structure, its secretion of hormones, and the activity of the enzymes that are necessary for steroid synthesis, the quality and quantity of the sperm produced and in the hormone dependent behaviour patterns of the male fish.

Testicular Structure

Endocrine disruption can cause changes in cellular structure or organisation that are specific to the endocrine tissues rather than due to non-specific cytotoxic action. Clear evidence for endocrine disruption of the testis is apparent when it steroid producing cells show increased or decreased activity the proportion of sperm at different stages of development differ from those unexposed fish or there is a complete arrest in sperm production. Such changes may be a result of either primary action on the testis or secondary action consequent upon action brain-pituitary complex.

Studies on 4-tert-butylphenol on common carp by A.V. Bares, et al (2006) suggested that carp testis loses it testicular architecture after exposure to 4-TBP. The reduction in spaces in the interstitium and lumen was distributed with disintegration of the lumen wall. Necrosis of the spermatozoa and also tissue fibrosis was observed along the patchy necrosis. These effect on the testis included reduction of testicular size and the reduction in number of germ cells. The increase in tissue vitellogenin concomitant with decrease in Gonado Somatic Index along with and elevation in Hepato Somatic

Indices of males treated with 4-TBP indicated that 4-TBP affected the reproductive organs directly as well as indirectly. Very adverse effect on the testis structure of young as well as sexually mature male carps exposed to 90-1000mg/L 4-tert-pentylphenol has been reported to cause inihibition of spermatogenesis, disappearance of spermatozoa and spermatogonic cyst (S. Gimeno *et al.*, 1998).

Xenobiotics as diverse as cadmium, pulp mill effluents, crude oil, PCBs and alkylphenolics all decreased concentration of the major testicular steroid 11-ketotestosterone in fish plasma, but it is often not clear whether such changes are cause or result of retarded development of the testis.

Brian. G. McAllister and D.E Kime (2003) reported that when zebrafish was exposed to tributylin (TBT) an aromatase inhibitor in different dose reproductive impairment was observed. Fishes exposed for 70 days to TBT with a dose range of lng/litre to 100 ng/litre showed abnormal sperm lacking sperm flagella and effect on sperm motility was pronounced. At lng/litre motility of sperm was significantly lower than that of control fish, while at l0ng/litre, all sperm lacked flagella, at 100 ng/litre milt volume had increased.

Gamete Viability

Although all male vertebrates produce vastly more sperm than eggs, evidence from both fish and mammals suggests that even a small decrease in sperm quality or quantity can decrease the males fertilizing ability. Viability of sperm in fish, as in other vertebrates is dependent on the correct hormonal and nutritional environment during their development within the testis. This may in turn be affected by the internal hormonal environment during early life stages in which the testis is differentiated. Endocrine disruption can therefore lead to abnormal development of the sperm and decrease its viability. This may take the form either abnormal sperm structure or decrease in its energy supply, both of which can alter its swimming ability and therefore its capacity to reach and fertilize the egg.

Additional Effect on Male

Successful courtship is dependent on the ability of the female to secret the pheromone, the male to detect it, and for his neural and endocrine system to elicit a response that both triggers his production of sperm and his appropriate sexual behaviour. Crabofuran and the organophosphate diazinon can both disrupt the ability of male fish to detect such pheromones (Moore A and Waring CP 1996).

Xenobiotics might also cause changes to other aspects of androgen-dependent breeding behaviour, particularly in species which male build or guards a nest or spawning territory like in hermaphrodite *Pseudolabrus sieboldi*. Development of male secondary sexual characteristics such as redbelly of sticklebacks or the hooked jaw in salmonids is dependent upon testicular secretion of 11-ketotestosterone, and any decrease in production of this steroid or effects on it receptor could inhibit his ability to attract a mate (Kime D E, 1999).

Vitellogenin (VTG) is a precursor of egg yolk that is synthesised in the liver of developing females under control of endogenous estradiol. In male, the circulating levels of endogenous estradiol are below detecTable limit, and there is normally no expression of the gene responsible for VTG production. If exposed to exogenous sources of estrogen, however, male fish are able to synthesise VTG in quantities approaching those of a female fish (J.P. Sumpter *et al.*, 1995). In common carp the vitellogenin in muscle homogenate was significantly different among groups when treated with 4-TBP (A.V. Varse *et al.*, 2006)

Strategy for Detection of Endocrine Disruption

If potential effects of endocrine disrupting chemicals are to be properly assessed, then it will be necessary to develop test system which are ecologically relevant, provide some means of gaining early warning of adverse effects, and link cause and effects. No single test system is likely to capable of fulfilling all these requirements. It is more likely that a multifaceted approach will be most successful. The following strategies are may be undertaken:

(I) Assay commonly used for screening compounds with endocrine disrupting activity includes ligand receptor binding assays, cell culture techniques, and measurement of estrogen mediated protein production. Physical and chemical fractionation methods rely on analytical chemistry to separate and isolate suspected EDCs from environment sample. Due to the wide variety of EDCs and other chemicals in the environment, these methods most suited to looking for known compounds frequently associated with endocrine disruption.

(II) Detecting endocrine disruption in situ and provision of early warning of potentially ecologically relevant effects might be achieved using biomarkers. Biomarkers are defined as biological responses that can be measured in tissues samples, excreta or at the level of whole organism, which signal exposure to or adverse effects of anthropogenic chemicals and radiation. Biomarkers which signal endocrine disruption have already been developed and there appears to be considerable potential for extension of this approach. For example, in a study by Purdom *et al.* (1994), exposure to domestic sewage resulted induction of vitellogenin, a biomarker of environmental estrogen exposure, in male rainbow trout. After only one week, vitellogenin level rose markedly. Subsequently vitellogenin induction has been used with other freshwater and marine fish species to signal exposure to xenoestrogens (Allen *et al.*, 1997). Another protein, egg shell protein or zona radiata protein has been shown to be catered 1000 times more responsive to xenoestrogens (Aruweke and Goksyor, 1996).

(III) A more precise form of environmental testing involves in vitro or single mode of action oriented in vivo tests (Baker, 2001). Most in vitro tests are hormone receptor binding assays, which test a compounds affinity for a given receptor, cell proliferation test, which measure a sample's ability to induce hormone dependent proliferation; or reporter gene assay, which use cells transfected with a reporter gene induced by a specific cellular receptor, providing a quantitative measure of receptor mediated gene induction (Villeneuve *et al.*, 2000).

(IV) DNA microarray-based technology, a new genomics technology is revolutionising our ability to achieve rapid throughput of thousands of simultaneous hybridisation reaction at a time. Significant development in robotics, surface chemistry and miniaturisation has permitted rapid developments of this technology. In most configurations, a DNA microarray is a glass microscope slide on to which many thousands of DNA samples have been spotted in a grid. DNA or, messenger RNA is extracted from cell or tissues, labelled with specific fluorescent molecules, and hybridised to the spotted DNA on the glass slide. The resulting image of fluorescent spots is visualised in a confocal scanner and digitised for quantitative analysis. There is tremendous potential in the application of DNA microarray technology to screen for EDCs. DNA microarray provide a snapshot of transitional activity in tissues samples showing which genes were actively expressed within the cell at one pint in time. By combining this technology with new teleost cell line grown in vitro, it should be possible to conduct relatively rapid straight forward assays to identify EDCs by observing the changes in gene expression pattern response to exposure.

Warning for Live Organism

Fish have many advantages as experimental models in the study of endocrine disruption, and although they do have some significant differences in their endocrine system to that of higher vertebrates and mammals, the underlying basis is very similar. Chemicals which are shown to be either actual or potential endocrine disruptors in fish can therefore certainly be considered as potential endocrine disruptors in mammals, including humans.

Fish thus provide an excellent biomoniotor of potential hazards of endocrine disruption to human exposure to similar chemicals. Although many xenobiotics bioconcentrate in the gonads, liver and other viscera, the edible muscle tissues still comprises the bulk of the fish and a high intake of dietary fish may lead to human exposure of high levels of potential endocrine disruptor. Such effects are most like to be seen in wild life and human communities with a diet rich in fish from contaminated water.

Future Needs

1. Current toxicity tests need to be improved with re-testing of substance undertaken.
2. Research needs to be adequately funded, prioritised and coordinated.
3. All India task force needs to be set up to assess the potential effects of hormone disrupting chemicals and opportunities to reduce their use.
4. Indian government should establish a unit or working group on endocrine disrupting chemicals.
5. Indian government should take immediate action to phase out and eliminate already identified endocrine disruptors. Furthermore endocrine disrupting properties should rank high under the prioritisation process for hazardous substance to be considered for such measures.
6. Research in both freshwater and marine fishes should be focused towards detecting hot spots for human exposure.

Although a good deal of information on nature and extent of comparative endocrine disruptor hypothesis has been generated over the past decade, more focused research is still needed to bring clarity on these issues.

Clear evidences have evolved that fish captured in polluted water suffer from a variety of endocrine disorders, including reduced fertility, abnormal sexual differentiation, developmental abnormalities and decreased response to stress. Such effects have been replicated in laboratory exposures and in *in vitro* experiments.

Similar problems have been found in fish eating birds and mammals, which have diet similar to those of fish eating human communities. There are increasing number of reports of decreased sperm counts and an increased incidence of genital abnormalities in human males.

While endocrine disruption in the human population will first become manifest at levels which do not attain the levels of clinical concern, even the slight decrease in fertility, response to stress and behavioural changes now apparent in fish and other wild life could threaten the cohesion of human society. Even though there is strong evidence that the danger exist, commercial and political vested interests are correct in their view that there is yet no absolutely conclusive proof that either is a threat to human health. If, however we wait until there is such definitive evidence it will be too late to take remedial action and inevitable consequences could cause catastrophic political and social upheaval.

3

Cryoconservation of Fish Gametes

Embryonic Cells

The archeological record supports five great extinctions since life began, each representing an evolutionary setback on the earth. The last of these mass extinctions happened some 65 million years ago, when the dinosaurs and at least one fifth of all other extinct life forms obliterated from earth.

Unlike previous ones which were caused by natural phenomenon, the present process of species extinction and natural disasters are of human origin. As various life forms disappear, causes changes in the earth's ecosystem diminishing the roles played by nature, such as pollution, seed dispersal, insect control and nutrient cycling. This loss of species is weakening the web of life and if it continues it would lead to irreversible and potentially unpredictable changes in the ecosystem.

Globally, the omens do not look good. Every where there are signs of the stretching of environment's limits. The global fish catch has now reached the limits of our fisheries. Stretching beyond the limits caused threat to fish may be the greatest of all, with nearly one third of all species—freshwater and salt water, now facing possible extinction.

Worldwide the principal causes of extinction are habitat loss and degradation, introduction of exotic species, over exploitation and other related factors. The main causes of habitat loss are agricultural activities, extraction (including mining, fishing and logging and harvesting) and development (human settlements, industry and associated infra structures). Habitat loss and fragmentation leads to the formation of isolated, small scattered population increasingly vulnerable to inbreeding depression, high larval mortality and susceptible to environmental stochasticity and consequently leads to extinction (IUCN, 2000). Over exploitation and other factors such as disease, pollution, genotoxicity and limited distribution are greatly responsible for endangerment and extinction of species.

Worldwide 11 per cent birds, 25 per cent mammals and 34 per cent of fish species are threatened (Lovell-Badge, 2001; Vemuganti and Balasubramanian, 2002) and aquatic ecosystem is becoming more vulnerable to disasters due to run-off from agriculture and industrial sources, oil spills or sudden environmental changes leading to total elimination of stocks from different ecosystems.

The underlying causes of biodiversity loss are macroeconomic policies, policy failures, poor environmental law/weak enforcement, unsustainable development projects and lack of local control over resources (Wood, *et al.*, 2000). About 24,600 living species of fishes in 482 families and 57 orders have valid scientific descriptions (Nelson, 1994). Approximately 23,000 species are bony fishes, 10,100 are entirely freshwater and 2500 moves between sea and freshwater during their life cycles. (Helfman *et al.*, 1997). The largest number of species occurs in the tropics and shows more biodiversity. An estimated 37 per cent of fish species that inhabit the lakes and streams of North America are either extinct or in jeopardy. Ten North American freshwater fish species have disappeared during last decades. About 68 per cent of native and endemic fish species have disappeared in semi-arid regions of Mexico, the situation may be even worse in Europe, where some 80 species of freshwater out of a total of 193 are threatened, endangered or of great concern. In South Africa two third of 94 fish species need special protection to avoid extinction.

India is ranked 2nd largest in aquatic biodiversity in Asia behind China (ICUN, 2000). The species diversity has been attributed to large river systems of 1.71 lakh km length comprise seven river systems and large wetlands (Bengal delta). Enhancing aquaculture production in India occupied more than 72 per cent share of total inland production (NFADP, 1999), thus relieved tremendous fishing pressure on open water systems. Still, rapid indiscriminate extraction of seed as well as brood fish from natural waters besides other anthropogenic activities caused continuous threats to native species. Captive rearing, unintended genetic manipulation, sex reversal, hybridisation and cross breeding already leads to genetic introgression and inbreeding depression. Release of fingerlings from such stocks into natural waters also contributed to indigenous species degradation.

It is urgently needed to take immediate measures to protect and conserve the species still that have the genes of economic, ecological and social importance. There are several options to conserve (protection/restoration of the ecosystem or native habitat of the species) or development of a breed to adapt changing environments and efforts should be prompted to create a need for products or functions of the breed is preferred as a mechanisms of conservation.

Unfortunately these are costly and require a grate deal of time, as habitat restoration *(in situ* conservation) is already a slow process. Other alternative is to maintain *ex situ* (conservation outside their natural environment) live or through cryopreservation of germplasm (gene banks).

Cryopreservation as a Disaster Management Tool

The tragic cyanide poisoning on the river Tisza (Hungary) in early February, 2000 (Horvath and Urbanyi, 2001) and mass mortality of fishes due to drop in the dissolved oxygen levels in lake Kanaria, Gujarat, India (Reuters, 2005) showed the vulnerability of the aquatic ecosystem to the world. In the event of such an emergency there is no method to restore a species in short time.

Death of genetically improved brood fish in the culture ponds which represents a static ecosystem similar to lakes causes considerable economic loss to the farmers worldwide. Sustainable aquaculture practices by judicious use of aquatic resources with minimal pollution are desirable, where endangered species find a suitable niche (Costa-Pierce, 1996: Costa-Pierce and Bridger, 2002). For that live gene banks are planned at many geographical localities as a conservation measure, but they are also susceptible to natural calamities. This emphasises the need of a cryopreservation protocol by which the germ cells such as spermatozoa of these dead fishes can be utilised to save species and prevent economic loss. Several highly effective cryopreservation protocols have been developed for mammalian oocytes and embryos (Hurst *et al.*, 1997; Karlsson *et al.*, 1996; O'Neil *et al.*, 1997).

Cryopreservation protocols based on data for mammalian embryos have been tested in isolated cells such as blastomeres (Leveroni and Maisse, 1998; Nilsson and Cloud, 1993; Strüssmann *et al.*, 1999) and whole eggs (Chao and Liao, 2001; Gwo, 2000; Hagedorn and Kleinhans, 2000) of fish, but successful cryopreservation was obtained only with the isolated cells. On the contrary, cryopreservation of spermatozoa is well established in a number fish species (Chao and Liao, 2001; Cabrita *et al.*, 2003; Suquet *et al.*, 2000; Babiak *et al.*, 1997; Horvath *et al.*, 2003; Lahnsteiner, 2000; Cloud, 2000; Gwo, 2000). The utilisation of fish cryopreserved sperm is finding its use in various aquaculture purposes such as stock upgradation in selective breeding programmes and as a source of gene banks in conservation of endangered species (David *et al.*, 2000; Mongkonpunya *et al.*, 2000; Tiersch *et al.*, 2000). It has been reported that the spermatozoa of fishes remain viable after the death (Koteeswaran and Pandian, 2002) and their cryopreservation and successful use to produce fish has been reported in *Labeo rohita* (Routray *et al.*, 2006). Nevertheless, this study has opened up new vistas in cryo-conservation of endangered fish and further studies must be conducted to produce androgenic population from the cryopreserved milt collected from dead fishes.

Principles of Cryopreservation

Cryopreservation means long term storage of biological materials without deterioration of its native characters over a long period at least several thousand years. The osmotic property of cells regulates most of the chemical events related with freezing. Intracellular and extracellular environments associated with water transport and membrane permeability form the basis behind cryopreservation of cells. Generally the larger cells that are less permeable should be frozen slowly than those that are smaller and more permeable should be frozen rapidly (Mazur, 1965).

The concept lies on the cellular damage during freezing process by osmotic shock, intracellular ice formation (IIF), increased intracellular concentration of solutes and solution effects (Mazur, 1977). Relatively large molecular weight of intracellular solutes generally does not passively cross the cell membrane leads to osmosis for maintaining osmotic equilibrium. The osmolality of a solution is a measurement of solute concentration and is a critical factor in the freezing process. This measurement defines the ambient environment of the cells under preservation and directs water movement across the cell membrane, which are crucial for the penetration of cryoprotectant into the cells during freezing procedures. The addition of cryoprotectants interacts with the membrane to make them more flexible and thus reduces damages due to solution effects (Mazur, 1977).

Ice Formation in Solution

Water with more solutes concentration freezes at lower temperature than pure water. In a standard solution, as pure water freezes out of suspension, leaving behind a solution of increased concentration. The freezing points of the remaining solution will therefore decline. Further decrease in temperature allows more water to remove in the form of ice and thus increases solute concentration further, which ultimately results in a continuous drop in the freezing point of the solution until reaches at the eutectic point, when entire solvent and its contained solutes freeze completely.

Super Cooling and Seeding

The concept of temperature changes and its effects on cells is prerequisite in understanding the cryopreservation process. In the process of cryopreservation under gradual decreasing temperature, water freezes out of solution and formed ice crystal until attainment of eutectic point. The supercooling, which is defined as the ability of liquid to cool several degrees below its own freezing point, yet still remain in liquid state. This is an unstable state that often results in spontaneous freezing and formation of large morphological structure of the ice crystals may then puncture the cell membrane and destroy cells (Mazur, 1963; Leibo, 1980).

The fluxing activities may be controlled to an extent during cooling procedure. The process of spontaneous freezing is refereed to as nucleation. To prevent the formation of the ice crystals, cells should be dehydrated as much as possible however the dehydration process only starts following nucleation. 'Seeding' (inducing ice formation in a controlled area of the sample) the freezing solution when it is just below its normal freezing point can induce IIF. Seeding process involves a pair of metal forceps cooled in LN2 and applied to the freezing container (straws) at a point of contact above the cell so as not to damage it in the freezing process.

Dehydration may also be achieved by decreasing the

eutectic point of the cell in solution by adding a cryoprotectant with a lower eutectic point. Rapid freezing may lead to the formation of lethal size ice crystals, but reducing solution effects, whereas slow freezing acts in reverse way. Large degree supercooling may be avoided by artificial seeding the sample at a temperature slightly below its freezing point.

Vitrification

Vitrification is the rapid cooling of liquid medium in the absence of ice crystal formation. The solution forms an amorphous glass as a result of rapid cooling by direct immersion of the embryos in a polyethylene (PE) straw into liquid nitrogen. The glass retains the normal molecular/ ionic distributions of a liquid but remains in an extremely viscous, supercooled form. The glass is devoid of all ice crystals, and the embryos are not subjected to the physical damage that is associated with ice crystal formation during freezing which is peculiar in conventional cryopreservation method. The cryopreservation techniques employ four basic steps such as addition of cell cryoprotectant, the freezing process (which involves cooling, seeding, freezing and storage in LN2), the thawing process and removal of the cryoprotectant (CPA).

Cryopreservation of Gametes in Fish

The cryopreservations of sperm have been reported from more than 200 species of fish with external fertilization. The present status of the art of cryopreservation for fish promises set up of gene banking. Selection of healthy ripe fish for collection of roe (milt) or eggs needs utmost care for better success in cryopreservation.

Time of milt samples is also important to obtain good quality sperm. Sperm samples collected at the beginning or end of the spermiation period may be of low quality, thus to be avoided (Billard and Cosson, 1992). As a practical recommendation, storage of untreated milt before cryopreservation should not exceed 1 hour, which may lead to lower fertilization capacity owing to changes in the composition of the seminal plasma and sperm metabolism (Lahnsteiner *et al.*, 1998 b). Thus initial steps of collection, quality control, dilution and freezing are to be completed with this time span of an hour. The spermatozoa counts vary species to species and even seasonal fluctuation also reported in many fishes. Sperm density reported to vary within a wide range of 0.8 x 10 9 cells per ml (CPM) for arctic char to 8.1 x 10 9 CPM for brown trout (Lahnsteiner, 2000).

The quality of sperm or milt required evaluation for better result, which is typically monitored by determining the number of motile sperm, the success of fertilization or by measuring some aspects of cell metabolism. Cryopreservation does not improve sperm quality, sperm motility and fertilization rate, rather may cause deterioration (Lahnsteiner *et al.*, 1996). Thus it is essential to assess the sperm quality, motility before preservation of sperm.

Though egg fertilization rate is best indicator of sperm quality, some physical measurement or either the whole sperm cell or selected components of sperm is generally used to predict fertilization rate. Sperm motility is frequently used for the purpose (Stoss and Holtz, 1983, Lahnsteiner *et al.*, 1995; Koupal *et al.*, 1995).

Measurement of the motility can be accompanied using simple microscopy or by CASA (Computer associated semen analysis). Though microscopy method is easy and rapid, but it is sometimes difficult to accurately assign motility values between 20 to 80 per cent. CASA with a high-speed video camera resulted in more accurate and descriptive determination of motility (Lahnsteiner *et al.*, 1996; Toth *et al.*, 1997). This technique provides information related to percentage of motile cells, types of movement and sperm velocity.

Flow cytometry has been successfully used to determine sperm quality in fish (Ogier de Baulny *et al.*, 1997). Measurement of a component within the sperm or seminal plasma also has been used to predict sperm quality (Baynes *et al.*, 1981; Lin and Dabrowski, 1996). Seminal plasma osmolality, pH and triglycerides were predictive of sperm quality. Osmolality showed wide variations among species and individuals. It is suggested that osmolality of media and diluents always be measured before use. It is observed that collecting milt into iso-osmotic diluent will improve sperm survival (Honeyfield and Krise, 2000).

The main role of seminal plasma is to create an optimal environment for storage of spermatozoa. Fish seminal plasma contains mainly mineral compounds and low concentration of organic substances. Three ions predominate in seminal plasma, sodium, potassium and chloride. Their concentration range between 75 and 175 mM for sodium, 32 and 86 mM for potassium and 112 and 183 mM for chloride (Marisawa, 1985; Sequet *et al.*, 1994). Calcium and magnesium ions also contribute significantly to ionic composition of seminal plasma. All these ions are involved in regulation of sperm motility either by their participation in sperm motility either by their direct or modulating effect on sperm activation (Scott and Baynes, 1980).

The first step of cryopreservation is the extension of cooled milt with solution containing cryoprotectants (CPAs) to allow the CPA to penetrate in to germ cells before freezing process. For successful cryopreservation, it is essential to prevent any activation of spermatozoa before or during freezing. Extenders composed of salts solution helps to prevent activation as well as in maintaining viability of cells during freezing process. Moreover use of extenders provides increased storage time. To maintain optimum sperm density below 2.0 to 2.5 CPM in storage container dilution with suitable diluents is necessary. Higher cell concentration limits intracellular space may cause damage, whereas lowers concentration (< 7.6 x 10^8 leads to inefficient loading of straws, which significantly influence the fertilization rate (Lahnsteiner *et al.*, 1996b)..

Two approaches have been followed to decide the composition of extenders. The first one involves a formulation mimicking the composition of seminal plasma, or use of physiological salt solution developed in fish (Stoss, 1983; Rana *et al.*, 1990). The second one involves formation of simplified extenders. Simple extenders vary often consisting of sugar and cryoprotectant, addition of cell culture medium have reported in some fish gamete cryopreservation.

Choice of any one by trial and error is the only option to develop suitable extender for a particular species considering an osmolality to maintain sperm cells in an inactivated state (e.g., 280 to 300 m Osmol/kg for freshwater species and 200-300 m Osmol/kg for marine species (Wayman and Tiersch, 2000). Most media feature a saline or saccharide bulk osmotic support, a suitable CPA at concentration varying from 0.2-1.5 M and various protective macromolucules additives or lipid components from vegetal origin.

In addition to seminal fluid and universal cryoprotectant, DMSO, other single CPA, combined CPAs, CPAs with additives have been used in cryopreserving gametes, blastomeres, and embryo. Pronounced species specificity has been observed in extender requirements. Glycerol, DMA (N, N-dimethyl acetamide), DMF (N,N-dimethyl formamide) and methanol have been used oftenly in fish species beside DMSO (Chao and Liao, 2001). Polyethylene glycol (PEG) and ethylene glycol (EG) are also used as CPA for their ability to rapid penetration and limited toxicity thereby producing less osmotic shock (Gwo, 2000). Extenders should be sterilised before use and antibiotics can be added to it to reduce microbial growth that reduces sperm viability (Stoss, 1983).

Dependant on dose, most CPAs are toxic to cells. To minimise toxic effects, they may be added slowly at cool temperature and should be diluted with extender solution prior to addition to sperm. Different dilution ratio of extender and CPA have been suggested which ranges from 1:1 to 1:20 for different species. The acceptable dilution ratio is 1:1 for grey mullet, black porgy and tilapia (Chao and Liao, 2001); 1:2 for artic char, 1:3 for white fish, Danube salmon, rainbow trout, brook trout (Lahnsteiner, 2000); 1:4 for milkfish; 1:5 for lake trout, 1:7 for brown trout, 1:9 for carp (Magyary *et al.*, 2000) and 1:20 for grouper, atlantic croacker (Gwo *et al.*, 1991)

Equilibration time is referred to the time needed for the cryoprotectant to penetrate the cells. Mostly it can be set at 15 to 30 min, but it may vary depending on the type and concentration of CPA being used and the type of cells under equilibration. Milt is generally cryopreserved with varied freezing protocols. Freezing protocols varies with species. Foe grey mullet, black porgy and grouper two step freezing by exposing straws to liquid nitrogen (LN_2) vapour at - 100°C and then quickly quenching to -196°C is optimal (Chao, 1991; Chao *et al.*, 1986, 1992). For carp, the best freezing protocol is -5°C/min from 2°C to -7°C and then -25°C/min from -7°C to -70°C and for tilapias, -20°C/min from 25°C directly to -35°C and then -5°C/min up to -70°C (Chao *et al.*, 1987), -12°C/min for Asian catfishes. -10 to -20°C/min for gilthead sea bream (Billard 1977b).

The two-step freezing method is simple to carry out and does not require equipment for controlled cooling rates. It appears that the cooling rate in the freezing phase has a very wide optimal range in the cryopreservation of sperm. Cooling rates vary with types of containers and different distances above the LN2 surface and thus different cooling rates can be achieved. Alternatively, a heated iron bar can form a series of tiny holes on dry ice (Leung and Jamieson, 1991). Large or small sized drops of milt mixture are dropped directly to those holes for cryopreservation in easy and farmer friendly way. Another choice is a programmable freezer, which is expensive to procure.

Thawing and Fertilization

Thawing temperature can have profound effects on sperm viability after freezing. Specific times and temperature should be optimised for thawing milt of particular species. A wide range of thawing temperature from 0°C to 60°C have been attempted to optimise protocols for each species. As a rule of thumb, samples are thawed when air bubbles within the straw can move freely within the liquid. If the straw is not transparent, the samples should be cool to the touch when thawed. The water temperature during the spawning season is usually ideal for sperm revitalisation. Rapid thawing of the sample, however will allow for rapid rehydration and markedly reduce ice damage. Thawing temperature of 40 to 50 °C for 0.5 ml straw equivalent to 7 sec works for many species. A serious problem occurs when CPA is removed from the cells.

To alleviate this problem cells are placed in a series of solution with decreasing concentration of CPA or in a extracellutar CPA to equilibrate. Thawed sperm should be added to eggs and thoroughly mixed and the gametes should be activated with an appropriate solution. At the post thaw phase, precautions should be taken such as, determination of sperm motility, shortening the time between thawing and fertilization and dilution rate. Since not all spermatozoa survive or maintain their fertility after cryopreservation, the density of the thawed sperm is preferably higher (almost 5 to 10 times) than that of fresh sperm to be obtaining maximum fertilization. The insemination ratio used may vary according to species and procedures between 10^4 to 10^7 spermatozoa per egg. The ratio of 2.5 to 5.5 x 10^6 spermatozoa per egg have been reported for Danube salmon, rainbow trout, lake trout and brown trout and 1.2 to 1.6 x 10^6/egg for gray ling and 3.0 to 5.0 x 10^5/egg for white fish (Lahnsteiner, 2000) and 1 to 1.5 x 10^5/egg for cyprinid fishes (Magyary *et al.*, 2000).

The fertilization and hatching rates obtained from cryo-milt were found equal or slightly lower compared with those obtained from fresh milt (Chao and Liao, 2001). Growth rates of hatched larvae did not differ significantly (Chao, 1991).

Oocytes

In the last decades, considerable progress has been made with cryopreservation of oocytes in a great number of species (Woelders *et al.*, 2003). Viable oocytes also have been recovered after freezing and thawing in some species. Post-thaw oocyte maturation, fertilization and embryo development have been reported in a number of species. Production of offspring from frozed thawed oocytes is still less efficient than that from cryopreserved embryos. Cryopreservation of fish oocytes is not successful (Blesbois *et al.,* 2003) due to large size, low permeability, sensitivity to chilling injury, high lipid content and the polar organisation (vegetal and animal pole).

Embryos and Embryonic Cells

In spite of a few claims (Wang *et al.,* 1987: Chen *et al.,* 1988: Zhang *et al.,* 1989; Zhang *et al.,* 1997), there is no reproducible cryopreservation protocols for fish embryo largely because of some limitations as in the case of fish oocutes, i.e., large size, the high lipid content and polar organisation in the fertilized eggs and the early embryos.

It has been reported that the yolk syncytial layer (YSL), a vitelline membrane formed by a single cell that surrounds the yolk and not the yolk or chorion embryos blocked the permeation of CPAs into the yolk atleast in Zebra fish (Zhang and Rawson, 1996b; Hagedon *et al.,* 1996; 1997b). An understanding of the physiology and removal of the permeability barrier could be a key to the future successful cryopreservation of fish embryos.

However cryopreservation of isolated embryonie cells in fish is an option. Post-thaw survival of blastomeres was achieved in rainbow trout, carp and medeka (Blesobois *et al.,* 2003). Embryonic cells can be used to produce chimeric embryos provided that the gonads of the recipient embryos are come implanted and populated with primordial germ cells (PGC s) from the donor embryo. Such chimeric embryo can be utilised to produce progeny of the donor genotype (Nakagawa *et al.,* 2002).

Somatic Cells

Cryopreservation of somatic cells was demonstrated for a number of cell types. Simple procedure of adding a suitable CPA to the suspended cell in culture medium followed by placing tubes at -80 °C in a mechanical freezer proved to be effective. Use of controlled rate freezer was reported in few cases.

Germ Cell Transplantation

Transplantation of ovarian tissues and germ cells (e.g. primordial germ cells, PGCs or spermatogonial stem cells, SSCs) are emerging technologies with potential for future conservation programme. Development of transgenic strains in a species of commercial importance and conservation needs time, labour and high financial supports, besides its unpredictable result. Germ cell transplantation is better option to overcome those difficulties.

The introduction of foreign gene in to the PGCs and selection of transformants facilitate the in vitro selection of transformants. The method is advantageous because of easy selection procedure of transformants, which can be performed in a petri-dish instead of rearing large number of fishes to identify the desired transgenic candidate and development of an individual fish using chimera technique is more convenient. Successful germ cell transplantation has been achieved in fish with fascinating results of allogenic transplantation in rainbow trout (Takeuchi *et al.*, 2003) leads to production of live fry with donor-derived phenotype from gametes of PGC-recipients.

4

Toxicology and Biomedical Research

Genetic engineers are inserting an increasing diversity of transgenes into an increasing diversity of aquatic animals (Kapuscinski, A.R. 2003; National Research Council. 2004). Genetically engineered organism' (GEO) and 'transgenic organism' refer to organisms bearing man-made recombinant deoxyribonucleic acid genetic constructs (Letourneau D. and Burrows B. 2002; National Research Council. 2002; Scientists' Working Group on Biosafety 1998). summarises representative examples of the growing diversity of fish and shellfish GEOs that are being engineered for aquaculture for a variety of purposes, e.g. for human food, for the biological control of nuisance species, for recreational markets, as water-quality monitors to detect contaminants that damage genes of living organisms, and even as bio-factories to produce commercially valuable compounds such as human Pharmaceuticals.

Growth-enhancement for human food production in aquaculture is the most common objective of current efforts but it may not remain so for much longer. Indeed, the first commercially marketed, genetically engineered animal in the United States of America (USA), and several other countries, was for the recreational, hobby aquarium market: the glow fish, a transgenic fish that 'glows' due to skeletal muscle expression of a fluorescent protein genetic construct (Food and Drug Administration. 2003; Gong , *et al.*, 2003). This growing diversity of objectives, species, transgenes and target traits suggests that future requests for commercial approval of aquatic GEOs will also involve a growing diversity of species, transgenes and target traits.

Historically fish have played in assessing potential risk associated with exposure to chemical contamination in aquatic environment, it is reasoned that the development of transgenic fish could enhance the role of fish in environmental toxicology and biomedical research.

Recent introduction of new models and methods demonstrates progress in application of transgenic fish in current research. Rapid advances are most evident in the area of *in vivo* mutagenesis using fish carrying transgenes that serve as recoverable mutational targets; these models highlight many advantages afforded by fish as a model and illustrate -important issues that apply broadly to transgenic fish to environmental and biomedical research.

Researchers have revealed that fish exhibit spontaneous frequencies of spontaneous mutation and respond to mutagenic exposure and consistent with known mutagenic mechanism. Results demonstrated that their feasibility of *in vivo* mutation analyses using transgenic fish and illustrated their potential value as a comparative animal model. Challenges to development and application of transgenic fish relate to need for improved efficiency in transgenic technology and in aspects of fish husbandry and basic research use. By taking advantage and of the valuable and unique attributes of fish as test organism. It is anticipated that transgenic fish make significant contributions for study in environmentally induced and genetically inherited diseases.

Transgenic Fish

With increasing numbers of synthetic chemicals introduced into the environment each year, concerns remain regarding our understanding of the linkages between exposure to toxic agents and potential disease. Although health effects associated with exposure to most chemicals are thought to be benign, actual risks posed by a majority of substances, particularly those entering ground water, streams, rivers, estuaries, and other aquatic systems, are largely unknown.

Chemical contamination of aquatic environments is of significant concern because although it is understood that aquatic systems serve as major conduits for distribution and deposition of many toxic agents, relatively few methods are available that provide sufficient sensitivity, accuracy, and practicality necessary for routine assessment of chemical toxicity. As a consequence, new approaches are needed to improve the assessment of health risks associated with exposure to chemical contaminants in the aquatic environment.

The development of transgenic animal models represents a revolutionary advance in the study of a variety of disease processes (Pinkert. 1997).

Using introduced genes, Endogenous genes, such as oncogenes or tumour suppressor genes can be selectively inactivated or knocked out to produce animals that do not express normal amounts of a specific gene product. Beginning in the late 1980s, researchers began to use genetically modified mice in toxicological studies of chemical carcinogenesis and in vivo mutagenesis (reviewed by Mahler 2000). Since then, the emphasis has continued to be on developing transgenic models to improve and expedite carcinogenicity testing (Battalora and Tennant 1999; Friedberg *et al.* 1998). Development of new transgenic rodent models for measuring other endpoints, such as xenobiotic metabolism and xenobiotic ligand /receptor interactions, is expanding recently in this field. (Mahler 2000). Considering the achievements of transgenic rodent models, it is reasoned that the development of transgenic fish could enhance the utilisation of fish as indicators of chemical exposure and as non- mammalian animal models in comparative biology. Fish have played historically significant roles in monitoring and assessing risks of exposure to chemicals in aquatic environments.

In some applications, such as assessment of toxicity associated with exposure to complex chemical mixtures or in low-dose chronic exposure regimens, fish are recognised as test organisms with distinct and superior benefits in providing insights to disease processes (Powers 1989).

Fish are also embraced as cost-effective and important animal models in genetics, developmental biology, and toxicology (Hawkins *et al.* 1995; Mullins *et al.* 1994; Solnica-Kretzel *et al.* 1994). Development of transgenic fish has been envisioned as a means to enhance the utility of fish models in reducing, refining, or replacing selected mammals used in toxicity testing.

In the following review, progress in the development of transgenic fish for studies of environmental toxicology is evaluated. Many advantages of fish as animal models are highlighted, and through an example of a transgenic fish model developed for in vivo mutagenesis, important issues are illustrated that apply widely to the use of transgenic fish in environmental toxicology.

Transgenic Fish Models

Scientists have generated numerous transgenic fish using a variety of species and transgenes since the first transgenic fish were introduced in 1985 (reviewed by Maclean 1998). Despite wide recognition of fish as sensitive subjects for chemical toxicity testing and as valuable comparative animal models, the field of environmental toxicology has only recently begun to benefit from transgenic technology.

The emphasis for transgenic fish research has been on refining transgenic methodology or developing novel strains of fish with commercially beneficial traits. However, recent developments indicate that transgenic fish are promising as new research animal models. To expand the utility of fish as environmental sentinel organisms (i.e., as indicators of the presence or biological impact of a hazardous substance), transgenic fish have been envisioned carrying reporter genes driven by promoters that are responsive to chemical exposure. In a generalised conceptual approach, the transgenic fish would be placed in the water containing the chemical to be tested. Following uptake, distribution, and accumulation of the substance in fish tissues, the genomically integrated response elements would be activated and the reporter genes would be up regulated.

The fish would be removed from the water and assayed for reporter gene activity that would be proportional to the concentration of the chemical to which the fish had been exposed. Using this approach, various transgenic fish have been conceived or are at early stages of development.

Examples include fish carrying metal-responsive or heat-shock promoters spliced into green fluorescent or *LUC* reporter genes to provide an in vivo indicator of exposure to heavy metals or other pollutants (Carvan *et al.* 2000; Gibbs *et al.* 1994; Maclean. 1998). Numerous groups have reported difficulties with sustaining gene expression in transgenic lineages, which has—slowed the introduction of new transgenic fish.

Expression of foreign DNA is commonly observed in the founder fish, but expression is rare or unpredictable in subsequent generations (Chen and Powers 1990; Fletcher and Davies 1991; Hackett 1993; Iyengar *et al.* 1996). Suboptimal expression may result from various factors including the use of heterologous transcription units derived from non- fish species (Foecking and Hofstetter 1986; Gorman *et al.* 1982) from methylation of CpG dinucleotides (Boyes and Bird 1991; Ehrlich *et al.* 1990), from incorporation of multiple transgenes into a single site (Amsterdam *et al.* 1995), or from integration of transgenes in loci susceptible to gene silencing (Eissenberg and Elgin 1991). As more groups gain experience using a variety of methods, such as insulating border elements to stabilise the expression of transgenes (Stief *et al.* 1989; Udvardy *et al.* 1985), it is anticipated that new fish models with improved transgene expression will be introduced.

A promising example is demonstrated by a transgenic fish assay recently introduced to determine the effects of estrogenic chemicals at critical life stages on sensitive target organs in fish. Legler and colleagues (2000) introduced an estrogen binding sequence linked to a TATA box and a luciferase reporter gene in zebrafish *(Danio rerio)* binding the chemical substance to endogenous estrogen receptors and subsequent transactivation of the estrogen receptors induced the luciferase gene as measured in tissue lysate. The study revealed that the period of gonad differentiation was highly sensitive in juvenile fish exposed to estradiol. In adult males, the testis was the target tissue most sensitive and responsive to estrogens.

Transgenic Fish Mutation Models

Researchers have developed mutation assays that are not reliant on the expression of a transgene in the animal to assess potential DNA damage after environmental chemical exposure. Using procedures similar to the ones Mirsalis and colleagues (1995) developed for rodents, Amanuma and colleagues 2000; (Winn *et al.*1995, 2000, 2001) produced several transgenic fish models that carry prokaryotic vectors harboring specific genes, which serve as targets for quantifying spontaneous and induced in vivo mutations.

The common approach to analysing mutations in transgenic rodent and fish models entails treating the animals with a chemical and allowing sufficient time for manifestation of the mutations. Genomic DNA is then isolated from various tissues and the vectors are separated and recovered upon transfer into specialised indicator bacteria where the mutant and non-mutant target genes are readily distinguished and quantified (Gossen *et al.* 1989; Summer *et al.* 1989). Transgenic mutation assays provide numerous benefits for analysing mutations in vivo not available using other approaches.

A significant challenge to detecting mutations in whole animals is the practical problem of recovering and identifying mutant genes. The need to detect mutations at very low frequencies (e.g., ~1 spontaneous mutation/105-107 loci) demands highly efficient means of recovering and distinguishing mutant genes among a very large number of non-mutant genes. Transgenic mutation assays afford efficient recovery and screening of large numbers of copies of a transgene target, providing statistically meaningful results as well as reducing the numbers of animals required in a chemical treatment.

Typically, hundreds of thousands of mutation target loci are recovered from as little as 1 to 5 ig of DNA, and as few as six to 10 animals/treatment are required to detect significant induction of mutations above a background mutation frequency (Boerrigter *et al.* 1995; Winn *et al.* 2000).

Transgenic mutation assays facilitate comparisons of mutational responses among identical loci recovered from different cells, tissues, organs, and species. In contrast to assays based on endogenous genes in which mutations are detected only in specific developmental stages or specific tissues, mutations in transgene targets can be examined in virtually any tissue from which DNA may be isolated. Mutation analyses can also be combined with measures of other endpoints such as sequencing of specific mutations to aid in disclosing possible mechanisms of mutagen action.

The utilisation of transgenic mutation models in a broad range of comparative analyses is improving our understanding of the influences of cell proliferation, metabolism, toxicity, and DNA repair in mutagenesis (Bielas and Heddle 2000). Before the advent of in vivo assays with the capability of detecting mutations directly in transgenic targets, researchers relied on genetic toxicity assays that focused on endpoints other than mutation. Among these endpoints, researchers have used induction of DNA adducts, DNA repair, DNA strand breakage, and chromosomal damage as indices of genotoxicity in vivo in fish (reviewed by Shugart 1995).

These analyses have provided valuable information regarding exposure of organisms to genotoxic agents and indirect information on alterations of DNA or chromosome structure induced by a chemical. However, these methods have limited sensitivity in detecting mutations at the level of the DNA, the ultimate endpoint of DNA damage and/or attempted repair.

Mutations in transgenic assays are detected in genetically neutral targets thereby avoiding potential selective pressures on the mutant frequency in vivo and allowing the accumulation and persistence of mutations (Cosentino and Heddle 1996; Swiger *et al.* 1999; Tao *et al.* 1993). As a consequence of the accumulation of mutations in these loci over time, repeated or chronic chemical treatments will increase the sensitivity of the mutation assay (Heddle *et al.* 1995).

The amenability of fish to a wide range of chemical treatment regimens indicates that fish will be well suited to mutation studies using treatments that closely approximate environmental exposure conditions.

Transgenic Fish Development and Care: Although development of new transgenic fish models has increased steadily inrecent years, transgenic fish are not currently being produced as routinely as rodents.

Production and perpetuation of lineages of transgenic fish remain a time consuming process requiring combinations of specialised personnel, equipment, and facilities that are not universally available in research laboratories. Procedures and facilities for developing and maintaining transgenic fish are nevertheless undergoing continual improvements to increase the efficiency of the process.

Choice of Species

Selection of a fish species for development of a transgenic model is dictated by whether the species has certain characteristics that will enable efficient production of the transgenic fish that can be used to address fundamental research questions. Small laboratory aquaria fish species such as medaka, mummichog, and zebrafish share numerous desirable traits for transgenic development and environmental toxicology. Small size, short generation time, and cost-effective husbandry contribute to efficient transgenic development and aquarium-based laboratory culture.

Well-described embryology, controlled year-round spawning, transparent chorion, and short embryogenesis and generation time are advantageous for transgenic production. Medaka has several characteristics that are especially well suited for environmental toxicology, including well-characterised histopathology. They have been used extensively in chemical hazard testing, carcinogenesis bioassays (Bunton 1999), and germ cell mutagenesis studies (Shima and Shimada 1994). As a common inhabitant of coastal environments, the mummichog is one of the most extensively characterised and utilised organisms in studies of marine environmental contaminant risk and eco toxicology (Lotrich 1975; Vogelbein *et al.* 1990).

Gene Transfer

Transgenic fish are produced through the introduction of foreign DNA sequences into zygotes or embryos resulting in the genomic integration of the DNA in a stable and heritable manner. Manual injection of DNA using drawn capillary needles via the cytoplasm of newly fertilized eggs remains the most commonly used method of introducing DNA into the fish genome (Chourrout *et al.* 1986; Maclean *et al.* 1987;Winn *et al.* 1995; Zhu *et al.* 1985). A variety of methods have also been used, including microinjection of the germinal vesicle (Ozato *et al.* 1986), electroporation of embryos (Inoue *et al.* 1990; Powers *et al.* 1992) or sperm (Muller *et al.* 1992), retroviral infection (Gaiano *et al.*

1996; Lin *et al.* 1994), and particle gun bombardment (Kolenikov *et al.* 1990). Additional methods will likely be introduced in the near future to exploit the genetic diversity offered by fish as a group more fully.

Establishing and Maintaining Transgenic Fish Lineages

Transmission of a transgene through the germ line is a requisite for most transgenic animal studies. Whereas the process of gene transfer can be technically challenging, the identification of transgenic founders, performance of appropriate crosses, and perpetuation of desired lineages are not technically difficult.

However, these tasks can burden resources and personnel. Mosaic integration of a transgene in founder animals (i.e., integration in only a portion of the animal's cells) is the rule in transgenic fish. Although the presence of the transgene may be demonstrated in excised fin tissue, the transgene is often not transmitted or is transmitted only at a low frequency to the next generation.

In addition, the need to identify transgenic fish that meet specific experimental requirements places additional demands on the process of establishing lineages. Problems may arise relating to integration or function of a transgene, or the animal may have poor reproduction or an undesirable phenotype thereby reducing the utility of the lineage. Consequently, it is advisable to establish multiple lineages to allow for these differences.

Fish researchers, being most familiar with the requirements for care, handling, and maintenance of laboratory fish populations, are typically the primary animal care providers. Unfortunately, fish culture facilities are generally not integrated into larger institution-wide animal resource programmes. As a result, fish researchers are often solely responsible for supporting their programme and meeting the demands associated with long-term perpetuation of wild-type and transgenic lineages.

These burdens can ultimately hamper production of new fish models and limit the availability of established lineages of animals for use by other researchers.

Similarly, other researchers may recognise the value of fish as models and may want to pursue the production of transgenic fish but may not have the necessary experience with fish husbandry or transgenic fish technology, further impeding introduction of new fish models. To assist in promoting optimal use of fish models and to bring standards, practices, and facilities for care and utilisation of fish up to that of mammalian models, various improvements are needed in aquatic animal resource Programmes.

In particular, improvements in training of personnel in fish care and transgenic techniques, increased research on diet, disease diagnosis and prevention, and increased Institutional support of aquatic research programmes will aid in expanding and developing new models and ensure continuity in long-term care. Before embarking on the development of transgenic fish, researchers are advised to address issues regarding assessment of the potential risks associated with environmental release of transgenic fish.

It may be argued that most small aquaria fish species are especially dependent on the laboratory environment for survival. For most transgenic models developed for applications in environmental toxicology, fish would conceivably carry new genes that are either neutral or deleterious to the animal's health. Nonetheless, it is advisable that procedures and facilities used to maintain transgenic fish be reviewed in accordance with guidelines proposed for the development, care, and use of genetically modified fish and shellfish to ensure that precautions are taken to safeguard against releases of transgenic fish into the environment.

Application of Transgenic Fish in Environmental Toxicology

Transgenic animal models hold significant promise for enhancing studies on the genetic and molecular basis of disease and for improving the scientific basis for environmental health risk assessment.

Studies from transgenic fish can yield reliable and reproducible data, using precisely defined toxicological endpoints while reducing costs and numbers of animals. By taking full advantage of the amenability of fish to a variety of manipulations, ranging from microinjection of embryos (Walker *et al.* 1996) to static-renewal and flow through chronic exposures (Kane *et al.* 1996), transgenic fish can improve the assessment of realistic risks related to exposure to waterborne and sediment-associated contaminants. As the utilisation of transgenic fish models intensifies, it is anticipated that transgenic fish will contribute significantly to emerging issues related to environmentally induced, reproductive, developmental, and heritable diseases for which such studies in mammalian models are exceedingly difficult.

The attraction of transgenic fish models is moderated by the uncertainty of how these new models will be applied most effectively in environmental toxicology.

Limited availability of lineages of fish beyond laboratories from which they were generated and limited databases on responses under a variety of test conditions will potentially slow wide adoption by researchers. For each new model and application, it will be necessary to resolve complex issues related to establishing optimal study designs, such as appropriate exposure regimens, chemical concentrations, routes and duration of exposures, and number of animals required for a test.

In addition, requirements for reproducibility of responses in individual test animals will demand rigid husbandry practices, many of which are not standardised among different laboratories, to reduce variability in culture conditions.

Guidance in addressing some of these issues may be obtained from following examples provided by application of transgenic rodent models in toxicology (Gulezian *et al.* 2000; Heddle *et al.* 2000; Mahler 2000).

Application of Transgenic Fish in Biomedical Research

The value of transgenic fish in investigating developmental processes had been well recognised and demonstrated in the mammalian and the and invertebrate models but the recent trends are coming up in the zebra fish as a human model for the various types of biomedical research and also development of knock out technology for many genetic diseases. The increasing popularity for the transgenic fish models has been gaining prominence and attributed to the successful application of Green fluorescent protein (GFP), also a reporter gene commonly called as a living colour fluorescent proteins Two fresh water fishes has proven (zebra fish, medaka) amenable for such transgenic analysis owing to their short generation of time, easy maintenance regular breeder, transparent embryos. The transgenic fish has increasing used for gene expression and function in development. Transgenic fishes are providing an excellent model for analysis of gene function in the *in vivo,* and *in vitro* systems.

The zebrafish is emerging as a powerful model system for cancer research. Several key aspects of the fish system make it particularly well suited to cancer gene discovery and to the genetic study of tumour biology in vivo. First, zebrafish exhibit true vertebrate cancer biology.

They are susceptible to spontaneous and carcinogen-induced tumours and the cancers they develop including carcinomas, sarcomas, haematopoietic neoplasms and germ-cell tumours closely resemble their human counterparts. Most of the known oncogenes, tumour-suppressor genes and apoptosis genes are present in the zebrafish genome and are highly conserved compared with mouse and human homologue. Second, the fish is a genetically tractable system. Fish can be cheaply maintained in large numbers; adults mate year-round and can produce clutches of several hundred embryos weekly.

Thus, large-scale forward genetic approaches are feasible.

Because the embryos are transparent and develop outside the mother, it is possible to investigate key processes such as cell proliferation and migration, apoptosis and angiogenesis in a living organism. Finally, an array of tools for the analysis and manipulation of gene expression now exists. An international effort has produced high-resolution physical and genetic maps, and a full genome sequence is nearing completion. Through transgenic approaches, oncogenes can be expressed in a tissue-specific or inducible manner. Morpholino oligonucleotides can be used to transiently knockdown expression of specific genes and high-throughput genomics techniques have been successfully used to generate animals inactivated for specific genes. Researchers interest has been in using the genetics of the zebrafish system to identify novel genes associated with susceptibility to cancer.

To create a zebrafish cancer model, screened the progeny of mutagenised zebrafish for mutations causing defects in embryonic cell proliferation. Screening fish at the embryonic level is rapid and efficient, and there is good reason to believe that mutations affecting cell proliferation in embryos will be relevant to cancer in adults (for example, inactivation of most tumour-suppressor genes in mice causes an embryonic phenotype). A screen using anti- phosphohistone H3 immunostaining as a proliferation marker yielded eight mutants, and we showed that the mutants have distinct and specific cell-cycle phenotypes.

These phenotypes include arrest or delay at specific stages of the cell cycle, aneuploidy, mitotic abnormalities and activation of DNA-damage checkpoints. All of the mutations are homozygous lethal and mutant embryos exhibit extensive apoptosis. A major question is whether adult fish with cell-proliferation mutations will have increased cancers. Because homozygotes have embryonic lethality, we developed a carcinogenesis assay to test for excess cancer susceptibility in adult heterozygote. There is a wide spectrum of tumours that develop after carcinogen exposure. These tumours closely resemble human tumours morphologically, and some lesions appear to progress through different stages of tumorigenesis.

In general, the cell-cycle mutants isolated via the genetic screen do not develop excess spontaneous tumours. However, in parallel to the cell-cycle screen and identified a family of zebrafish that develops into a spontaneous tumours at high penetrance, with dominant inheritance, and as early as three months of age. Interestingly, more than 90 per cent of these tumours are testicular germ cell tumours (seminomas). Little is known about the molecular pathways causing seminomas in humans. To gain insight into this disease, cloning was done to the mutant gene. Through interval haplotype analysis havé mapped the germ cell tumour mutant to a 20cMinterval on zebrafish chromosome 10, and continue to collect more affected animals. The early success of the zebrafish system in generating transgenic cancer models, inactivating key genes and identifying cancer-related genes is encouraging. The next step is likely to come from combinations of these approaches. For example, expressing oncogenes in the context of an underlying cell proliferation mutation may provide new insight into cancer-related pathways. A particularly exciting prospect is the further use of large-scale screens. For example, given a phenotype due to the loss of a tumour-suppressor gene, a second site mutation that suppresses the phenotype could identify molecúles interacting with the tumour-suppressors. These molecules might be excellent therapeutic targets. This intersection of classical genetics with vertebrate cancer biology may allow the fish system to make major contributions to our understanding of cancer biology.

Conclusion

In conclusion, the increasing number of substances introduced into the environment each year, combined with the prevalent need for improved methods to assess potential adverse health effects, will be an impetus to continue to develop new transgenic fish models. The speed with which these new models can be developed and the extent to which they will be adopted will be influenced largely by advances in transgenic technology and improvements in fish care and use.

Table: Examples of genetically engineered fish under development Species Target engineered traits proposed application Status of development Finfish

Mud loach	Increased growth rates, Improved feed conversion after insertion of construct containing growth hormone gene driven by a strong promoter; construct is a novel recombination of mud-loach genes (Nam. 2001; Nam 2000)
Channel catfish	Enhanced bacterial resistance after insertion of moth peptide antibiotic, cecropin B gene (Dunham, 2002)
Grass carp	Increased resistance to grass carp hemorrhage virus after insertion of human lactoferrin gene (Zhong 2002)
Medaka	Transgenic fish serve as a detector of mutations (presumably caused; by pollutants) that could affect aquatic animal or human health, a method has been patented After insertion of mutagenic bacteriophage vector, vector deoxyribonucleic acid (DNA) is removed and inserted into indicator bacteria to measure mutant genes (Winn) R.N. 2001; Winn R.N.,1995; Winn R.N. 2000; Winn R.N.2001)
Atlantic salmon	Increased growth rate and food conversion efficiency by inserting. Chinook salmon growth hormone gene and antifreeze gene Federal Department of promoter. (Cook *et al.*, 2000; Hew C.L. and FletcherG.L. 1996).
Zebra fish	Fluorescent red or green body colour (Gong, 2003) Hobby aquarium market

	Red sea increased growth rates after insertion of ocean pout antifreeze protein bream gene promoter and Chinook salmon growth hormone (Zhang *et al.*,1998)
Rainbow trout	Improved carbohydrate metabolism after insertion of human glucose transporter type I or rat hexokinase type II genes driven by viral or piscine industrial uses promoters. Potentially allows higher plant-material content in fish feeds (Pitkanen *et al.*, 1999) Trout Increased growth rate and food conversion efficiency via insertion of sockeye Model transgenic fish line. Salmon growth hormone gene (Devlin *et al.*, 2001) for public-domain research
Carp and medaka	Goldfish Production of male-only offspring by insertion of gene constructs that prevents Biological control of the fish's aromatase enzyme from transforming reproductive hormone androgen nuisance species, such asinto oestrogen; to prevent development of female fish (Thresher R. & Bax N. 2003)
Goldfish	Increased cold tolerance after insertion of ocean pout antifreeze protein gene (Wang *et al.*, 1995)
Tilapia	Increased growth rate and food conversion efficiency after insertion of tilapia preparing to seek regulatory growth hormone gene (Martinez *et al.*, 2000) approval. tilapia Production of clotting factor after insertion of human gene for clotting Pharmaceutical

	production Research is ongoing factor VII, for medicinal applications (Aquagene L.L.C. 2003).
Tilapia	Increased growth rate, food conversion efficiency, and utilisation of protein after insertion of Chinook salmon growth hormone (Rahman 2001).

Ornamental Fish as Model Animals for Biotechnological Research

Fishes are of immense value to mankind as food, feed, fertilizer, medicine and recréation (sport fisheries). Aquaculture and capture fishery provide employment to millions, either directly or indirectly. There are several fish species of recreational and aesthetic value.

Ornamental fishes form the most ideal group under this. These fishes have proved to be useful also as scientific model for biomedical research, studying molecular mechanism of evolutionary adaptation, monitoring of aquatic pollution and for analysing the mechanism of tumour formation (Schartl, 1995; Holder and McMohan, 1996; Ishikawa *et al.* 1997). For instance, puffer fish, a strongly-poisonous marine ornamental fish, is used for isolation of unknown vertebrate genes for its small and compact genome (Brenner *et al.*, 1993).

Aquaculture is one of the fastest growing commercial sectors globally. The global market for tropical fish has boomed in recent times particularly due to ornamental fish. Compared with growth rates of 3 per cent for livestock meat and 1.6 per cent for capture fisheries, commercial aquaculture has registered an annual growth of about 10 per cent since 1984 (NACA/ FAO, 2000). Ornamental fish are the most valuable fishery commodity in the world on an unit weight basis and its export value was approximately $US 350 million worldwide as by the turn of the century (Hardy, 2003).

Over the past decade, genetic engineering has emerged as one of the most powerful transforming technologies known and this along with conventional biotechnology is staged to rule the world over the current decade. It is possible now to transfer the beneficial traits of a particular gene from one organism to another in far less time and precision than what was needed in traditional breeding.

The technologies have embarked on the mainstream research so much, so that ethical and legal issues have mushroomed all over the world, such as the 'Issues in Science and Regulation of Transgenic Fish'. This describes some of the products that could be created through the application of genetic engineering to aquaculture as well as the potential environmental and food safety hazards/issues associated with such products.

Fish in general and ornamental fish in particular have umpteen uses in biotechnological research and are regarded as scientific model, attributable mostly to two reasons:

First, they produce eggs in large quantities (high fecundity) and these in turn develop outside the fish body, with a minimum survival rate of 40-50 per cent. In contrast, other test animals such as the rats, rabbits, guinea pigs and other such mammalian conventional test organisms have low fecundity that too remaining inside the mother's (brood's) body. Once scientists genetically engineer the somatic or growing egg, it must be inserted back into the animal.

Second, an equally compelling reason is that ornamental fishes are less sturdy than their edible counterparts, susceptible to diseases and sensitive to pollution. These characteristic features of ornamental fishes make them useful experimental tool in various biological researches, from organism to cell to genome to gene level.

The various fields where ornamental fish has a role as biotechnological tool range from bioassay, pharmaco-toxicology, eco-toxicology, toxico-genomics, toxico-proteomics, pathology, parasitology (host-pathogen interactions including *in vitro* interactive-enzymology), proteomics, genomics, phylogenetics, agriculture, germplasm conservation (cryopreservation), molecular characterisation to probiotics to elaborate a few of the state-of-art research *modus operandi*.

Ornamental Fishes as Biotechnological Tool in Disease Study

Zebra fish, *Brachydanio rerio* is used as major test organism in biotechnological study. Its transparent embryos are ideal to study the link between development and immunity. Zebrafish embryos already possess immune-competent macrophages after one day of development. Genome analysis shows that 90 per cent of all human genes have orthologs in zebrafish. In addition to *in vivo* studies, we use various zebrafish and human cell lines for signal transduction research.

Cancer Study

Body organs of zebrafish have many similarities to humans such as the brain, heart, liver and pancreas, and contract diseases like cancer. A comparison of the human genome sequence and the soon to be completed zebrafish sequence demonstrates conservation of cell-cycle genes, tumour suppressors, and oncogenes. Beyond comparative genomics, there are many advantages to modelling cancer in the zebrafish system.

Cell-lines study on effect of onco-drugs on as a part of the pharmacogenomics and toxicogenomics studies were undergone. Most importantly, via the carcinogenesis assay, the zebrafish system provides a direct way to test, if a mutation causing an embryonic phenotype also causes a cancer predisposition in adults. Transgenic lines could be used to express oncogenes, or to delineate gene expression through the use of green fluorescent protein (GFP). Beginning with the pioneering work of Streisinger and colleagues, the zebrafish was envisioned as an excellent model system for complex biology (Amatruda *et al.*, 2002).

ZF-MODELS/Zebrafish models were made for human development and disease study (EU 6th framework). In this European project microarrays and proteomics are being used to aid the study of signalling pathways induced by glycan molecules that are involved in the communication between cells in relation to pathogen recognition, development and cancer. The signalling functions of glycans that represent ligands of the Toll-like receptors, such as hyaluronan oligosaccharides, were studied.

Apoptosis Study

The study encompassing a chain of (cytopathic) events that ultimately leads to the death of a healthy cell after exposure to any adverse parameter forms the part of apoptosis (programmed cell death). Apoptosis is important in normal biological processes and in pathogenesis in vertebrates. In a study, caspases and other apoptosis-regulating genes have been cloned from zebrafish *(Brachydanio rerio)* and other fish species. Elucidation of in vivo functions of apoptosis is focused on development, morphogenesis and sex differentiation. In an attempt to elucidate cause and effect relationships between caspase and development, transgenic zebrafish overexpressing procaspase-3 were generated. Stress-induced apoptosis in zebrafish embryos can be monitored by whole mount TUNEL staining and caspase assay. Thus, zebrafish is a useful experimental model animal for investigation of apoptosis *in vivo* (Yamashita, 2003).

Study has demonstrated that UV-inactivated Largemouth Bass Virus (LMB V) induced apoptosis in Bluegill fry-2 (BF-2) infected cell monolayers via the extrinsic (caspase-8) pathway (Treadaway, 2006). Addition of pokeweed anti-viral protein (PAP) to virus infected cells at a concentration sufficient to inhibit viral replication failed to inhibit LMBV-induced apoptosis. Viral protein synthesis was necessary to reinitiate apoptosis via the intrinsic (caspase-9) pathway. Activation of the caspase-9 pathway might have been in response to the virus-induced apoptosis. The figure below discusses modern approaches in pathological studies that include finding origin and answer at protein and DNA level.

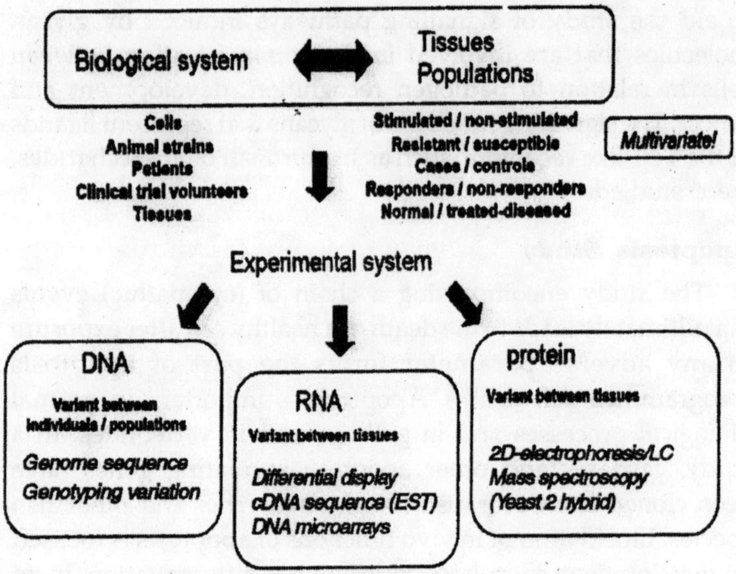

Fig. Linking biological variations to disease pathophysiology

Ornamental Fishes as Biotechnological Tool in Etiology

Comparative characterisation, pathogenicity, biochemical, genomic and serological studies showed nodavirus infection in various isolates of diseased guppy, *Poicelia reticulata* (Hegde *et al.*, 2003).

The biotechnological tools/methods involved in the study are worth-mentioning. The nucleotides of various strains were compared.

Western blot analysis using rabbit antisera raised against the nodavirus from marine fish, *Epinephelus tauvina* confirmed its antigenic similarity to the marine nodavirus isolate. Asymptomatic infection in guppy fry was observed following experimental infection with this virus and the marine nodavirus isolate (Singapore strain) implying the spread of virus from marine fish to freshwater fish. Thus, by use of above methods first nodavirus infection in freshwater fish was confirmed.

Various morphological studies identified and confirmed a

capillarid nematode infection in zebrafish kept at several research facilities and in a large carcinogen exposure study previously conducted at Oregon State University.

The worm was identified as *Pseudocapillaria tomentosa,* a common nematode of cyprinid and other fishes. Capillarid nematodes may have intermediate or paratenic hosts. A laboratory transmission study confirmed that the parasite has a direct life cycle (Kent *et al.,* 2002).

Nightlights for the Aquarium Used as Research Tools

Genetic tinkering has resulted in fluorescent fish and a boost for Taiwanese ornamental fish dealers. These can also be used as GM house pets and known as Night Pearl. The technology for the breakthrough was based on the research of Tsai Huai-jen of the National Taiwan University (NTU).

In 2001, he announced creation of fluorescent fish and saw the potential to sell them as aquarium fish. The genetically-modified trait of spectral glow by the otherwise ordinary-looking Japanese medaka and zebra fish made these livelier for the aquarium tank.

To achieve this, the scientists extracted fluorescent-protein-encoding-genes from jellyfish and corals and injected them into the fish at the embryo stage. Though work on using fluorescent fish as research tools are going on for some time now, only Singapore and Taiwan have some success story to share till date.

Ornamental Fish as Bioreactors'

At present the most promising application of transgenic technology in animals is to use transgenic animals as a 'bioreactor' to produce useful proteins and compounds in medicine and other areas. This is also known as 'gene farming' where gene of other animal or species can be inserted in a specific tissue so that it produces useful substances. Current biotechnological research efforts are not limited to improving farm productivity.

In an attempt to identify new sources for production of pharmaceutical substances, scientists have engineered tilapia with a human gene that makes the fish capable of producing a compound essential to the production of Factor VII, a human blood-clotting protein.

Scientists have tried to improve cold tolerance of goldfish by engineering hsp (heat-shock protein) genes so that it thrive cutting across global agro-climatic zones, than the wild counterpart (Hwang *et al.*, 2004).

Scientists have altered some marine microorganisms to reduce their dependence on light so they might live in unique dark environments, as is at the bottom of deep sea and ocean.

These microbes have also found a place in aquatic medicines, the area of research focusing on the anatomy, histology, immunology, pathology, diseases and therapeutics of commercially cultured finfish species including food-fish (such as trout, catfish, tilapia, flounder, yellow perch and striped bass), ornamental fish, bait fish and tropical fish.

The advantages of the transgenic fish bioreactor include the speed of generation of transgenic fish, low cost, and low risk of transmitting animal pathogens, etc. Furthermore, the fish muscle is edible and raw meat may be directly consumed for better preservation of bioactivity; thus, nutritional and pharmaceutical/therapeutical supplements could be expressed in the fish muscle system.

For edible vaccine, there should be a great potential in aquaculture as edible vaccine fish can be produced in feed fish and be used to feed other aquaculture species to reduce the laborious works of immunisation of huge number of individuals.

To increase the yield of the fish muscle bioreactor system, fast-growing and large size farm fish such as carp, tilapia, catfish, salmon and rainbow trout may be used. For all these species, the transgenic technology is well developed (Gong and Hew, 1995).

Comparison of protein content in hen egg white and genetically modified fish

Animal	*Gestation months*	*Maturation months*	*Recombinant protein yield<10g/l*	*Elapsed months from microinjection to protein*
Hen-egg white	0.7	6	1 egg/day or 1 mg/l	6
Fish	0	3-12	27 g/kg	<4

Ornamental Fishes as Tools in Environmental Biotechnology

Ornamental fish are being used as test organism for toxicity study through bioassay. This focus includes the sub-disciplines of neurotoxicology, immunotoxicology and environmental toxicology. Ornamental fishes are exposed to increasing problem of aquatic pollution and many species can only be saved through breeding programmes. Several studies have documented heavy metal toxicity on growth uptake and elimination of metals (Cuvin and Furness, 1988), reproduction (Kaviraj, 1983; Nicoletto and Hendricks, 1988) and metal removal by biochemical agents (James *et al.*, 2000) in cultivable fishes. However, toxic effects of heavy metal on growth and reproduction in ornamental fishes were not studied by many workers. Ornamental fishes are not sturdy like edible fishes and they are easily susceptible to diseases and pollution. The findings of copper toxicity on aquarium fish could be applicable to cultivable fishes too. A study was undertaken to study the toxic impact of copper on growth and reproductive performance in sword tail, *Xiphophorus helleri*. The major effects were vertebral deformity, low reproductive potential and growth reduction.

Some of the other recent terms that one invariably encounters while studying bioassay and toxicity are genotoxicity, toxicogenomics, cytogenetics, pharmacogenomics, etc. Genotoxicity assays, for instance, are carried out on the basis of the parameters, such as, micronucleus test, chromosome aberration studies, sister-chromatid exchanges and single-cell gel electrophoresis. Use of these has increased significantly in the past few years in, for example, detection of damage due to genotoxicity, radiation, oxidative damage, apoptosis, etc. Numerous studies conducted at the Station have led to improved understanding of phosphorus and selenium metabolism and patterns of phosphorus excretion in rainbow trout and cut throat.

A semi-static bioassay was used to study acute toxicity of endosulfan to zebrafish *(Brachydanio rerio)* and yellow tetra *(Hyphessobrycan bifasciatus)* which included behavioural changes like hyperactivity, erratic swimming and convulsions observed at the highest exposure levels.

Histological studies showed predominant acute effects in the gills, with inflammatory infiltration necrosis and separation of respiratory gill epithelium (Jonsson *et al.*, 1993). Studies were undertaken on the influences of two commonly used organophosphate pesticides, methyl parathion and phosphamidon, on the reproductive activities of female spotted Munia *(Lonchura punctulata)* (Maitra *et al.*, 2000). Researchers are also assessing the feasibility of fluorescent fish as water quality/safety indicators.

GFP transgenic medaka line with an estrogen responsible promoter derived from the *vitellogenin (vtg)* gene has been generated and characterised. It was demonstrated that GFP expression could also be induced by several other synthetic and environmental estrogenic compounds, including 17-a-ethynylestradiol (EE2), diethylstibestrol (DES), estriol, and bisphenol-A (BPA), indicating the broad application of the transgenic line to monitor a wide range of estrogenic endocrine disrupters (Zeng *et al.*, 2005b).

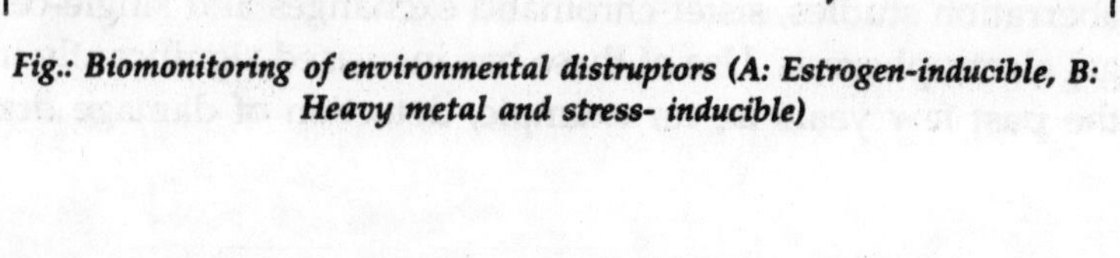

Fig.: Biomonitoring of environmental distruptors (A: Estrogen-inducible, B: Heavy metal and stress- inducible)

Ornamental Fishes as Tools in Molecular Biology

As of recent times, the studies on molecular biology include various sub-disciplines including vaccine development, transgenics, molecular parasitology, molecular virology and molecular mechanisms of infectious disease, developmental biology, gene expression, transcription, signal transduction, etc. Molecular-level manipulation by employing basic principles of molecular genetics and allied techniques could find solutions for increasing the overall fish production. The molecular level studies could be used in several areas such as population genetics, brood stock development, fish health management, control of reproductive cycles, transgenics, genetic diversity and conservation, fish genomics, etc. In fisheries biology, molecular markers can be used as valuable tools in systematics, phylogenetics, brood stock management, population structure, species identification, gene mapping, selective breeding, QTL mapping, legal applications and many others.

GM fish offer new potential for increased production of cultured organisms. This technology allows the introduction of new traits or the improvement of the old ones in a way that is impossible to achieve with conventional breeding methods. Examples of genes with commercial potential are those which control growth, disease, cold tolerance, sexual maturation and meat quality/preservation.

The Omics Revloution (Proteomics and Genomics)

The omics revolution of science include functional genomics, structural genomics, physiological genomics, toxicogenomics, proteomics, metabolomics, etc. Genomics involves the study and understanding of genes and their implementation in the development of health care products and services.

A genome include all the genetic information encoded in a cell involving the study of the gene which is the chromosomal segment that codes for a single functional polypeptide chain or RNA molecule. While proteomics is a new science that focuses on the study of proteins: their roles, their structures, their localisation, their interactions, and other factors. Proteomics analyses, for example, the proteins of human fat cells, corn leaves, or an organism like the bacteria. A recent proteomics study has shown that wild and farmed fish have distinctly different protein profiles, although the nutritional implications are unclear at present.

Other current projects include using proteomics to diagnose the effects of nutrition and environmental fluctuations on the proteins in fish, including the effects of endocrine disrupting compounds. A third omic which is different from the above two, involve analysis of total metabolite pool ("metabolome") to reveal novel aspects of cellular metabolism and global regulation is metabolomics.

These omics concepts have been yielding results in various pathophysiological studies and their potentials are just limited by human imagination.

Ornamental Fishes as Biotechnological Tool in Proteomics

Fish cell-lines: Fish cell-lines have been used to support many areas of research, such as, virology, immunology, genetics, environmental toxicology, aquaculture and seafood quality. This rapid growth was facilitated by the ease with which cell-lines are cultured, maintained and manipulated. They grow well at room temperature without the need for heat-controlled incubators or carbon dioxide buffering as is a case with cell-lines of higher vertebrates; their low metabolic rates translate into little maintenance over a long period of time.

Among the various techniques for detection of such cross-contamination, DNA profiling, PCR methods, use of antigens and proteomics techniques are notable. And now proteomics has been applied to the fish cell-line cross-contamination problem by Lucy Lee and Sarah Wagg from the Department of Biology at Wilfrid Laurier University in Waterloo, Canada. The approach based on the comparison of protein expression profiles that avoids the identification of individual proteins which can be difficult because genomic and proteomic databases do not hold data on many fish species.

This new research relies on the comparison of proteins that are at relatively high concentrations, so the risk of misidentifications/ miscomparisons is very low.

The protein expression signatures clearly differentiated between the different cell lines but also clustered together those originating from the same species.

The profiles originating from trout and salmon, two closely related species, matched more closely than those of more distantly related species like trout and goldfish or eel. So, qualitative differences between the proteomes of different cell lines can be discerned.

The method is designed specifically to achieve a rapid and reliable comparison of cell lines and will be in reach of many labs because it uses basic, comparatively cheap instrumentation and does not require specialist techniques like mass spectrometry.

Animal Proteomics

Proteasomes are large, multi-subunit particles that act as the proteolytic machinery for most of the regulated intracellular protein degradation in eukaryotic cells. An alteration of proteasome function may be important for the regulation of the meiotic cell cycle.

To study the change at the subunit level comparative proteome analysis of changes in the 26S proteasome during oocyte maturation in goldfish was performed Two-dimensional polyacrylamide gel electrophoresis was used to separate proteins.

For differential analysis, whole spots of the 26S proteasome from goldfish oocytes were identified by matrix-assisted laser desorption/ionisation-time of flight mass spectrometry and database analysis. These results demonstrate that modifications of proteasomal subunits and cell cycle phase-dependent interactions of proteins with proteasomes occur during oocyte maturation in goldfish (Horiguchi. *et al.,* 2006).

Protein Markers in Phylogenetic Study of Ornamental Fishes

Genetic variation will ultimately tend to proteomic polymorphism within limited time span and morphological differences in broad spectrum of time frame.

The similarity in protein banding pattern between Goldfish, *Carassius auratus* and Koi Carp, *Cyprinus carpio,* may be attributed to the biochemical and metabolic correspondence among the fishes. Although differences among species in enzyme maximal activity or concentration are often interpreted as adaptive and important for regulating metabolism, these differences may simply reflect phylogenetic divergence. The results indicate similar metabolic requirements of these enzymes in the particular organs. Various biotechnologically useful enzymes and proteins can be compared and characterised in this phylogenetic study.

The protein and enzyme concentrations also indicate phylogenetic similarity and divergence with goldfish and koi carp being more closely related in these aspects than to gourami or texas. Obtained data can influence the understanding the digestive processes of investigated fish. Thus it can be elicited that enzyme is the physiological marker of the body and is the reflection of genetic variation (Malik, 2006).

Ornamental Fishes as Biotechnological Tool in Genomics

Genomics involves the study and understanding of genes and their implementation in the development of health care products and services. Genomics involve philosophical (new way to approach biological problems) and a practical (high-throughput methods for analysing biological systems) approach and is a fast evolving field. Genomics is being used for developing improved fish varieties for aquaculture.

Ornamental Fish in Studying Host-parasite Interactions through Comparative Cellulase Enzyme Estimation

The proteomic and host-parasite interaction pattern as an integrated aspect of polymorphism study of important ornamental fishes can be investigated through both qualitative and quantitative enzymatic activities in fish gut. Cellulase enzyme producing bacterial flora isolated from fish digestive tracts could be exploited in bioconversion of recalcitrant lignocellulosic wastes into useful products such as glucose fructose syrup, single cell protein cultivation and as fermenTable sugars to alcohol.

On seeing the above economical uses of cellulase, extracting it from fish gut bacteria could be beneficial, subject to economic feasibility of the total process. Molecular biology of host-parasite interactions inside the fish intestine could be investigated. This necessitates phylogenetic reconstructions for both hosts and parasites. Comparison of these phylogenies will be used to test hypotheses concerning co-speciation and co-evolution in biogeographical contexts.

In due course of time this microorganism evolved in the same way as fish does due to similar environment. Thus it can be used to study fish phylogenetics and conservation of pure lines. *In-vitro* culture of cellulase producing microorganism by providing conditions similar as that in the host's intestine, will certainly enhance cellulase yield as the microorganisms will flourish owing to host-parasite relationship.

Thus, further molecular study on this work is needed to get better knowledge of this relationship which can also be used in understanding and implementing it in other areas such as natural biocosmetics, pest traps, etc. and in cellulase production and extraction which is used in manufacturing bioethanol.

On studying this work at molecular level a putative transgene can be introduced in the microorganism with a promoter which enhances cellulase production. This, after adaptation in intestine and establishing host-parasite relationship will be further helpful in increase in cellulase production and such studies will thus be possible. In the present scenario where non-renewable energy resources are on the threshold of depletion, this will be a big boon for mankind.

Singapore grouper iridovirus (SGIV) genome was compared with other five iridovirus genomes at the DNA and protein levels. Proteomic analysis showed evidence of novel proteins detected at the post-transcriptional level. Studies gave important information on molecular mechanism of virus-host interactions and will have a broad impact on future strategies for the design of specific inhibitors or drugs to control these pathogens in genera.

Ornamental Fish in Germplasm Conservation

Fish variety gets hybridised with undesirable morphologically similar variety. This hybridisation, either naturally, or directed by man, has widely propitiated the contamination of cultured stocks. Thus, conservation of ornamental fish germplasm is the need of the hour. For the study of phylogenetic variation in fishes, various scientists have differentiated them on the basis of morphological, molecular and biogeographical patterns.

Ornamental Fish in Molecular Characterisation and Phylogeny

The main sources of markers are mitochondrial DNA (mtDNA), minisatellites, microsatellites, introns, and anonymous nuclear sequences assayed using highly specific PCR primers or by using the RAPD technique. A mushrooming market of ornamental fish in general and goldfish in particular is in the offing (Bandyopadhya *et al.* 2004). Goldfish being a Cyprinid (a modified variety of *Cyprinus carpio)* could be used as an ideal model for biotechnological studies in larval and juvenile cyprinids. It has huge commercial value both in foreign and domestic market.

Due to its commercial importance and high reproductive ability, it is necessary to develop an improved strain. Some of the efforts have been taken place to assess the fish germplasm (Bardakci and Skibinski, 1994). A phylogenetic profile is a pattern of presence or absence of a particular gene or protein across a set of organisms.

Proteins that function in the same cellular context frequently have similar phylogenetic profiles which have proven to be helpful in the elucidation of the evolutionary history of biochemical pathways or protein machineries. The fast growing number of completely sequenced genomes continuously enhances the resolution of phylogenetic profiles. The studies shows that DNA markers are useful tools for estimating inbreeding depression and heterosis in guppy *(Poecilia reticulata)* breeding (Shikano and Taniguchi, 2003). Protein as well as DNA-RAPD studies showed that Goldfish is phylogenetically more related to koi than any other ornamental fishes among goldfish, koi, texas and gourami. Phylogenetically Gourami was most distant to goldfish (Malik, 2006). Zebrafish are surface swimmers in conditions of broad and shortwave-dominated background spectra and goldfish are generalised swimmers whose light environment extends to a depth of elevated short wavelength absorbance with turbidity. Studies indicate that the evolutionary engineering approach is very effective in deciphering the process of functional divergence of visual pigments (Chinen *et al.*, 2005).

Ornamental Fish as Tool for Trobiotics' Research

A termed coined and first used in aquaculture about two-decades before by Jory *et al.* (1986), probiotics (pro-favour; bios-living beings) unlike antibiotics refer to organisms that favour the growth of a target organism. 'Probiotics' generally includes bacteria, cyanobacteria, microalgae and fungi, etc. referring to all those microbes which could improve general water quality inhibiting the water-borne pathogens, thereby increasing productivity.

Probiotics increase the population of fish-food organisms; improve the nutritional levels of aquacultural animals and their immunity to pathogens. Sugita and Shibuga (1996) isolated bacteria from the intestine of 7 kinds of freshwater cultured fish and reported their anti-bacterial abilities. They concluded that the presence of intestinal bacteria protected the fish against the infection by pathogens.

Researchers at Heriot-Watt University have developed a range of probiotic microorganisms which show protection against fish diseases such as furunculosis, enteric redmouth disease, vibriosis and winter ulcer disease. In addition these organisms provide a dietary supplement to enhance the growth, nutrient utilisation and overall health of carp species (Swain *et al.* 1996).

Ornamental Fish as Tool in Reproductive Biology

The Reproductive Behaviour of Asian Arowana: Pioneering research collaboration is on with Qian Hu Fish Farm on the reproductive biology of a primitive teleost Asian arowana (*Scleropages formosus,* Osteoglossidae) or dragonfish. Asian arowanas are interesting for basic research due to their unusual reproductive strategy (a few eggs of huge size protected by mouth brooding) and they command an extremely high price in the ornamental fish trade as these are regarded as most favourite omen in *'Feng Sui'*. Their partner selection process is being studied by using genotyping with polymorphic DNA markers isolated in the lab and trying to develop methods to predict mate choice on the basis of genotypes.

Genomic Analysis of Teleost Sex Chromosomes: The majority of the 24 thousand fish species do not seem to have highly differentiated sex chromosomes and those that do seem to show tremendous variability. Differences among male and female genomes are being searched by PCR-based methods during the past ten years and DNA markers associated with sex from three different teleost species (African catfish, Asian arowana and turbot) have been successfully isolated.

Sex-associated DNA markers might be used for increasing the density of genetic linkage map of sex chromosomes. The comparative analysis of sex chromosomes will help to understand the secrets of their evolution. Sex-specific DNA markers allow developing "molecular sexing methods", which are useful for improving the aquaculture production of farm fish species with differential growth rate and maturation time between the two sexes.

Comparative Genomic Study of Gonad Differentiation in Cyprinid Teleost

Genes with conserved regulatory roles in the gonad differentiation of common carp (*Cyprinus carpio* L., Cyprinidae) and zebrafish are being searched in the genomic library. By using testis samples from all-male common carp populations generated through androgenesis, cross-species hybridisation experiments with the microarray are being performed. Candidate genes with putative role in the gonad differentiation in teleosts have been studied.

Tool in Mosquito Control

Gambusia affinis is the most commonly used biological control agent for mosquitoes in the world and are known as mosquito fish. Mosquito-fish are important species in research on fish reproductive physiology and environmental toxicology. Careful use of the fish can provide safe, effective, and persistent suppression of a variety of mosquito species in many types of mosquito sources.

Many species of larvivorous fish have been evaluated as agents to control mosquitoes, including various species of atherinids, centrarchids, cichlids, cyprinids, gasterosteids, and other poecilids. Mosquito-fish have certain characteristics that make them efficient predators of mosquito larvae. They thrive in shallow warm environment in water where mosquitoes prefer to lay eggs. These are surface-oriented predators where mosquito larvae are an accessible prey. Small size enable them to penetrate moderately penetrated and shallow areas within the mosquito source. In many sources, seasonal peaks in mosquito-fish activity and population growth coincide with mosquito reproduction times. Because of their omnivorous feeding habits, mosquito-fish can thrive in habitats where mosquitoes occur intermittently. Being hardy these are easy to handle, transport and stock. Mosquito-fish provide inexpensive and safe long-term control, sometimes within days after application.

Population genetic structure of the western mosquito-fish, *Gambusia affinis* in San Antonio river in San Antonio was studied using allozyme electrophoresis. These results suggest that selective forces such as non-point source runoff or low dissolved oxygen, perhaps in combination with limited migration due to numerous dams, have reduced genetic diversity of populations in the downtown area (Roark *et al.*, 2001).

Studies were undertaken to characterise multiple forms of vitellogenin (Vg) in mosquitofish (*Gambusia affinis)* by using methods of column chromatography, full length cDNA cloning and sequencing of Vgs and to discover the fate of each Vg during its processing into product yolk proteins.

These developments set a stage for elucidation of the detailed physiological mechanisms underlying nutrient supply to embryos of viviparous fishes, which remain largely unexplored.

Ornamental Fish for Hybridisation Aberration Studies- Mosaic Fish

Over the past decade, genetic engineering has emerged as one of the most powerful transforming technologies known to this generation. Scientists can now transfer the beneficial traits of a particular gene from one organism to another in far less time than needed in traditional breeding and with more precision. In aquaculture, the most successful application of the transgenic technology should be the generation of growth enhanced *GH* transgenic fish.

It was observed that cytoplasm affects the number of vertebrae in Carp-goldfish clones. Although the initial rate of success in producing carp-goldfish clones is low, seven offspring in 501 attempts in this study, the authors believe that cross-species transplantation will lead to improved understanding of the contributions of the nucleus and egg cytoplasm to the growth and development of vertebrates (Sun *et al.*, 2005).

Ornamental fish can also be used as biotechnological tool for developing mosaic fish. Colour gene was derived by two tissue specific promoters, one for skin cells because of the exterior location and the other for muscle cells because of the large mass. These promoters on ligation to the jellyfish *GFP* gene generated stable transgenic zebrafish lines using both chimeric genes. Another colourful transgenic zebrafish was developed, using the same muscle specific *mylz2* promoter with RFP (red fluorescent protein) and YFP (yellow fluorescent protein) genes (Gong *et al.*, 2003).

To investigate the possibility of creating more transgenic colours, double transgenic fish were obtained by crossing a *mylz2:GFP* fish with an *mylz2:RFP* fish. The double transgenic fish displayed orange fluorescent colour.

It was found that by mixing two muscle extracts respectively from the *mylz2:GFP* fish and *mylz2:RFP* fish at different proportions, a series of intermediate fluorescent colours from green, yellow, orange, to red can be created.

Thus, the number of colours used for transgenic ornamental fish will not be limited by few available living colour genes if promoters of different strengths are used to express fluorescent proteins to different levels or proportions.

A fluorescent transgenic fish are being marketed in United States with the trademark GloFishTM as the first transgenic animal product. It is expected that GloFish will sever as an excellent biotechnology model for public education to understand transgenic technology, GMOs (genetically modified organisms) and GM products. In deed, GloFish is probably so far the best illustrated transgenic model for general public because the easily observed traits changed by genetic modification and the amusing appearance of these fish.

Tool in Ontogeny

The transgenic technology is becoming increasingly popular as a powerful experimental tool in developmental biology of model fish, notably the zebrafish *(Brachydanio rerio)* and medaka *(Oryzias latipes)*. From late 1980s to late 1990s, the research in this area is basically for technology development. With the demonstration of faithful expression of the *GFP (green fluorescent protein)* living colour reporter gene under homologous tissue specific promoters (Higashijima, *et al.*, 1997; Ju *et al.*, 1999), the power of the *GFP* transgenic fish has been quickly recognised.

Now the *GFP* transgenic fish system is actively used to investigate gene expression patterns, tissue/organ development, tissue-specific promoters/enhancers, cell lineage and migration, upstream regulatory genes, mutagenesis screening and characterisation, promoter/enhancer, chimeric embryos and fish cloning by nuclear transplantation, etc. (Gong *et al.*, 2001, 2004).

In the past few years several tissue-specific promoters have been isolated and characterised from zebrafish and to develop dozens of stable *GFP* transgenic lines. These transgenic lines generally express transgenic GFP in tissue-specific patterns essentially identical to the expression patterns of endogenous genes from which the promoters derived.

Thus, these transgenic lines provide valuable tools for to analyse development of certain organs and tissues. For example, the fluorescent reporter genes under the fast muscle-specific *mylz2* promoter are expressed exclusively in fast muscle in the trunk and head regions, identical to the pattern of the endogenous *mylzl* gene that codes for a fast muscle isoform of myosin light chain 2 proteins (Xu *et al.*, 2000; Ju *et al.*, 2003). Promoter *mylz2* is also used to drive other living colour reporter genes in transgenic zebrafish, including BFP (blue florescent protein), RFP (red fluorescent transgenic protein), and YFP (yellow fluorescent protei.

Similarly, Other tissue-specific expression of living colour proteins were achieved by skin epithelium-specific *keratin8 (krt8)* promoter (Gong *et al.*, 2002), oocyte-specific *ZP3* promoter and exocrine-pancreas specific *elastaseA (elaA)* promoter. Meanwhile two two-colour transgenic lines have also been developed, one with *mylz2* promoter to drive *RFP* and *krt8* promoter to drive *GFP* and the other with *ela*A promoter to drive *GFP* and the liver-specific *L-FABP (liver fatty acid binding protein)* to drive *RFP*.

Table. Summary of stable lines of living colour transgenic zebrafish and medaka developed in the laboratory

Model fish	Promoters	Living colour gene expression	Reference
Zebrafish	*krt8:GFP*	Skin and intestinal epithelia	Gonq *et al.*, (2002)
	krt8:GFP)/mylz2:RFP	Two-cclour transgenic zebrafish with GFP expression in skin epithelia and RFP in muscle	Wan *et al.*, (2002)
	mylz2:GFP	Fast skeletal muscle	Ju *et al.*, (2003)
	mylz2:YFP	*ibidem*	Gong *et al.*, (2003)
	mylz2:BFP	*ibidem*	Pan and Gong, unpublished
	*mylz2: zs*Green	*ibidem*	Pan and Gong, unpublished
	Mylz2:loxpGFP	*Ibidem* but with loxP sited flanking GFP gene	Pan *et al.* (2005)
	elastaseA	exocrine pancreas	Wan *et al.* (2005)
	Zp3:GFP	oocytes	Liu *et al.*, MS under preparation
	L-FABP:RFP/elaA:GFP	RFP in liver and GFP in exocrine pancreas	Pan *et al.* unpublished
	Hsp25:GFP	Multiple tissues and inducible by heat shock	Pan *et al.*. manuscript in preparation
	Vimentin:GFP	Neurons	Pan *et al.* unpublished
Medaka	Zebrafish *mylz2*	Muscle	Zenq *et al.* (2005a)
	Medaka *mvtg1(GFP)/zebrarish keratin8(RFP)*	GFP expression in female liver and estrogen inducible in both sexes. RFP expression in skin and intestinal epithelia	Zengefa *et al.* (2005b)

Advantages of Using Ornamental Fish as Biotechnological Tool: By using biotechnological techniques, including molecular and recombinant technology, scientists study the growth and development of ornamental fish to understand the biological basis of traits such as growth rate, disease resistance or resistance to destructive environmental conditions.

Researchers are using biotechnology to identify and combine valuable traits in distant-parental fish to increase productivity and improve product quality. The traits scientists and companies are investigating for possible incorporation into several organisms include increased production of natural fish growth factors and the natural defence compounds organisms use to fight microbial infections. Biotechnology is also improving productivity through the development of feed additives, vaccines and other pharmaceutical agents.

Ornamental fishes are small, easy to handle, can be studied and maintained in the aquariums as test animals. An aquarium makes an ideal ecosystem to study interaction of fishes with their surroundings. In environmental biomonitoring, though all the animals are affected by pollution but only fishes are taken as model as these are susceptible to diseases and sensitive to pollution.

The benefits from such transgenic fish are obvious; e.g. increase of yield, shortening of the production time, decrease of operation cost, potentially lowering the market price for consumers' benefit, etc.

A commitment to research and to development in ornamental biotechnology will respond to the critical needs of society by developing novel drugs for treating diseases, providing new techniques to monitor and assess, restore , protect and monitor environments, ensure sustainable and safe aquaculture and fisheries, discovering new types of composite materials, biopolymers and enzymes for industry.

Disadvantages of Using Ornamental Fish as Biotechnological Tool: Current problems of marketing transgenic fish are environmental concern (not limited to GM fish, *e.g.*, exotic species, etc.), food safety, ethical concerns, gene flow (horizontal gene transfer and food chain), vector DNA (antibiotic gene, viral DNA such as promoter and poly A signal, DNA replication site, transposon, etc.), viability of transgenic fish is no better than that of the wild type, fecundity of transgenic fish is much lower than that of wild type fish, fertility of transgenic fish is slightly lower than that of wild type fish, transgenic fish usually stop growing after 4-6 months while wild type fish can continue to grow to a larger size, transgenic fish are disadvantageous in mating selection, there is no reproductive advantage of transgenic fish in term of the number of transgenic offspring, no indication that fluorescent transgenic zebrafish pose a higher risk than the wild type fish to ecosystem/environment. However, because of concerns in food safety and ecology, none of the growth hormone transgenic fish has so far been approved for commercialisation.

Biotechnological Research

The application of genetic engineering to animals could provide numerous benefits, including the possibility of a safer, cheaper food supply and the creation of new sources for scarce pharmaceutical resources.

But, as demonstrated by the current plan to regulate transgenic fish, it is unclear if regulators have the tools they need to adequately evaluate these new products, and there is reason to wonder if innovation is getting ahead of the ability to manage it.

Without a clearly articulated and transparent road map to guide agency review and approval, it will remain difficult for developers to bring products to market and even more difficult to assure the public that the next generation of agricultural biotechnology products are undergoing a regulatory review process that includes a careful, thorough, and rational consideration of risks and benefits.

Though ethical committee has put many restrictions and legislations behind the use of other animals like guinea pigs, rats, etc., they have not kept any such limitations on the use of fishes so that these can be smoothly used as research model.

In pollution control mosquito fish are not effective in all situations. Since these are not stocked in numbers sufficient to cause an immediate effect, they do not control mosquitoes as quickly as pesticides do.

In some areas, federal, state, or local agency permission is required to stock mosquito fish. In recent years, scientists have applied genetic engineering for a wide variety of purposes to a host of animals traditionally used as food sources, including cows, pigs, and fish. Only one animal— genetically modified fish—is thought likely to come to market in the near future.

Conclusion

Ornamental fish are not sturdy like edible fish and they are easily susceptible to diseases and pollution. Thus, these could prove to be useful tools for their biological behaviour and ecological/organismal interactions. Improving production efficiency could lower cost and provide economic benefits to producers and consumers.

However, global supply and demand combined with consumer acceptance will largely determine whether these benefits will be realised in terms of economic gains. Ornamental fish are the most valuable fisheries commodity in the world today in terms of returns per unit area.

Further, owing to the fact that these are such suitable specimen from conventional to the most modern type of biological research, some of the concerns that need proper and immediate address include, survival of existing species; maintaining the potential of species for continual evolution; utilitarian value; collection of baseline data on ecosystem, socio-economy and diversity of the population in order to analyse the impact of the ornamental fish trade on social and natural environments; expanding their biodiversity and genetic characterisation studies. The 'ranching' works as carried out by ZSI and other such Governmental and non-governmental bodies may further be encouraged.

To summarise, ornamental fishes could prove beneficial in disease study, bioassay, toxicology study, pathogen study, proteomics, genomics, Host-pathogen interactions in phylogenetics study, *in vitro* cellulase culture (economical cellulase extraction) study, agriculture, germplasm conservation, molecular characterisation and phylogeny study, as also probiotics and other such multi-disciplinary applied research. The following table summarises the various products from fishes being used as biotechnological tools for carrying out numerous research studies and having commercial applications.

Table. Summary of various products from fishes and their uses

Source	*USES*
Fish	Cosmetics, tissue replacement, anti-coagulant
Fish	Prevention of heart disease, mental development in premature children, anti-tumoural lipid metabolism
Fish hydrolysates	Antioxidants, immunostimulants, Nutraceutical product
Zebrafish, Medaka	Living colour qene expression
Zebrafish	To study cancer
Zebrafish	To aid the study of signalling pathways human development and disease
Zebrafish	To express oncogenes, or to delineate gene expression through the use of green fluorescent protein (GFP)
Zebrafish	To elucidate cause and effect relationships between caspase and development
Fish	Knowledge of viral functions, such as DNA replication and transcription, nucleotide metabolism, protein processing, manipulation of cellular responses and virus-host interaction medaka and zebra
Fish	Genetically-modified trait of spectral glow
Zebrafish, Medaka, Goldfish, etc.	Production of Factor VII, improvement of cold tolerance of goldfish by engineering hsp; nutritional and pharmaceutical/ therapeutical supplements expressed in the fish muscle system
Medaka	Broad application to monitor a wide range of estrogenic endocrine disrupters
Zebra fish. Yellow tetra, Munia, rainbow trout, Cutthooat trout, and ornamental fish, *Swordtaill*	Monitor heavy metal toxicity on growth; to study phosphorus and selenium metabolism and patterns of phosphorus excretion, toxicity of endosulfan, methyl parathion and phosphamidon
Goldfish, Eel, etc.	To study cross-contamination, qualitative differences between the proteomes of different cell lines can be discerned
Zebrafish	Amusing appearance, excellent biotechnology model for public education to understand transqenic technology
Carp and Goldfish	Understanding of the contributions of the nucleus and egg cytoplasm to the growth and development of vertebrates
Zebrafish	To investigate gene expression patterns, tissue/organ development, tissue-specific Dromoters/enhancers, cell lineage and migration, upstream regulatory genes, mutagenesis screening and characterisation, promoter/enhancer.

Fish in general and ornamental fish in particular have an added advantage as biotechnological tool specifically in animal biotechnology due to the fact that fishes as yet do not convincingly come under the new *'Animal Ethics Code'* as being implemented by the Department of Biotechnology as a part of the code of conduct for all the institutes involved in carrying out animal biotechnological research, including preclinical trials and animal metabolism studies.

Thus, ornamental fishes are better than conventional methods used to study and implement as new biotechnological tools in various fields.

5

Environmental Biotechnology and Agriculture

Biofertilizer

Biofertiliser (*jibaja sar*) living materials used to increase fertility of soils. Some free-living or symbiotic bacteria and blue-green algae (Cyanobacteria) fix gaseous nitrogen as ammonia and release it increasing the fertility of soil and water. *Rhizobium* or *Bradyrhizobium* producing root nodules in legumes and *Anabaena azollae* living in leaf cavities of *Azolla* (aquatic fern) are very efficient nitrogen fixers, and contribute about 500 kg N/ha/year.

Another microsymbiont with nitrogen fixing capacity is *Frankia* having hyphal morphology similar to that of actinomycetes and produces nodules in woody non-legumes, like *Alnus, Casuarina, Myrica,* etc. It produces vesicles and sporangia in both free-living and symbiotic states. *Azotobacter* species are free-living (mostly root associated), aerobically nitrogen fixing bacteria. *Nostoc, Calothrix, Gloeotrichia, Stigonema,* etc. are free-living aerobically nitrogen fixing Cyanobacteria. In addition, Vesicular Arbuscular Mycorhizae (VAM fungi) are free-living soil forms that increase nutrient uptake (specially by converting organic phosphorus into inorganic phosphorus), plant growth, nodulation and nitrogen fixation in legumes. In coastal areas of some countries, seaweeds are also used as biofertilisers.

However, all these life forms may be grown artificially and inoculated in the soil as biofertiliser. The nitrogen fixers releases nitrogen during their life time and also add other elements after their death and decay, essential for the growth of crops. The biofertiliser also contributes organic matter and maintains a good soil texture. *Azolla* contains well balanced amino acids and high amount of anthocyanine, a -carotene. These compounds also have positive effects on growth and yield of crops.

Large-scale production (2.5 m tons/year) of *Rhizobium* or *Bradyrhizobium* inoculants has been successful at Bangladesh Institute of Nuclear Agriculture, Mymemsingh. About 1.5 to 2.0 kg inoculant/ha used with seeds may result increased production of pulses by 20 to 40 per cent. Average *Azolla* production in a pond could be up to one m ton/ha/day, following the method developed at Botany Department, Dhaka University. It could also be grown along with irrigated rice crop. One cover of *Azolla* is about 10 m tons/ha and incorporation of two such covers could reduce urea-N fertiliser by about 50 per cent. *Azolla* could also be made into compost and addition of about 10 m tons could supplement all fertilisers by about 50 per cent.

Impact on Agriculture

The unprecedented growth in crop yields and agricultural productivity over the 20th century owes much to a series of biological innovations embodied in seeds, beginning with the development of hybrid crops in the United States in the early part of the century and continuing with high-yielding varieties during the Green Revolution of the 1960s and 1970s. More recently, developments in modern biotechnology are expanding the processes of biological innovations by providing new tools. Agricultural biotechnology is a collection of scientific techniques, including genetic engineering, that are used to create, improve, or modify plants, animals, and microorganisms.

Genetic engineering (GE) techniques allow a more precise and time-saving alteration of a plant's traits (facilitating the development of characteristics not possible through traditional plant breeding), and permit targeting of a single plant trait (decreasing the number of unintended characteristics that may occur with traditional breeding). Despite the benefits, however, environmental and consumer concerns currently limit acceptance of agricultural biotechnology, particularly in Europe. The ultimate contributions of agricultural biotechnology will depend on our ability to recognise its potential benefits and its risks (Fernandez-Cornejo *et al.*, 1999).

Despite a focus here on genetically engineered crops in agriculture, the future importance of genetically engineered animals should not be understated. As a National Research Council (NRC) report indicates, the increased demand for meat and deterioration and loss of agricultural land will lead to pressures to exploit biotechnology to improve productivity in animal agriculture.

GE crops are often classified into one of three generations (Panos, 1998). First-generation crops have enhanced input traits, such as herbicide tolerance, insect resistance, and resistance to environmental stresses like drought. Second-generation crops have added-value output traits, such as nutrient-enhanced seeds for feed. Third-generation crops produce pharmaceuticals, bio-based fuels, and products beyond traditional food and fibre. At present, GE crops widely adopted are first-generation.

For the most part, farmers have adopted herbicide-tolerant (HT) varieties, which help control weeds, faster than insect-resistant varieties.

Weeds are such a pervasive pest for soybeans, corn, and cotton that over 90 per cent of planted acreage for each crop was treated with herbicides in recent years. Acreage share for HT soybeans has expanded more rapidly than that for HT varieties of cotton and corn, reaching 87 per cent of US soybean acreage in 2005.

Farmers' adoption of HT soybeans has been widespread among major growing States, ranging in 2005 from 76 per cent in Michigan to 95 per cent in South Dakota. Acreage share for HT cotton has also expanded rapidly, reaching 61 per cent in 2005. In contrast, acreage share for HT corn reached only 26 per cent in 2005, but this has also trended upward since 2001 (Fernandez-Cornejo, 2004b).

Insect-resistant crops contain a gene from a soil bacterium, *Bacillus thuringiensis* (Bt), which produces a protein toxic to specific insects. Acreage shares for Bt cotton and corn are lower than those for HT soybeans and cotton and vary much more across producing States, with adoption more concentrated in areas with high infestations of targeted pests (insect infestation varies much more widely across locations than does weed infestation).

Farmers planted Bt cotton to control tobacco budworm, bollworm, and pink bollworm on 52 per cent of cotton acreage in 2005. Acreage share ranged from 13 per cent in California to 86 per cent in Louisiana. Bt corn, originally developed to control the European corn borer, was planted on 35 per cent of corn acreage in 2005, up from 29 per cent in 2003 and 24 per cent in 2002. The recent increases in acreage share may be largely due to the commercial introduction in 2003/04 of a new Bt corn variety that is resistant to the corn rootworm, a pest that may be even more destructive to corn yields than the European corn borer.

Other GE crops used by US farmers over the past 10 years include herbicide-tolerant canola, Bt potatoes (introduced by Monsanto in 1996 and withdrawn from the market after the 2001 season), virus-resistant papaya (developed by Cornell

University and University of Hawaii and introduced commercially in 1998), and virus-resistant squash. In addition, a tomato genetically engineered to remain on the vine longer and ripen to full flavour after harvest was introduced by Calgene in 1994, but withdrawn after being available sporadically for several years (Colorado State University, 2004).

Main Reasons Stated by U.S. Farmers for Adopting GE Crops

According to surveys conducted by USDA in 2001-03, most farmers (59-79 per cent) adopting GE corn, cotton, and soybeans indicated that they did so mainly to "increase yields through improved pest control". The second most cited aim was to "save management time and make other practices easier" (15 to 26 per cent, except for Bt corn, which was much lower); the third reason was to "to decrease pesticide costs" (9-17 per cent of adopters). All other reasons combined accounted for 3-7 per cent of adopters. Hence, factors expected to increase economic profitability by increasing revenues per acre (yield times price of the crop) or reducing costs (operator labour, pesticides) are expected to promote adoption most.

Adoption of GE Crops and Yields

The first generation of GE crops does not increase the yield potential of a hybrid. In fact, yield potential may even decrease if the varieties used to carry the herbicide-tolerant or insect-resistant genes are not the highest yielding cultivars. However, by protecting the plant from certain pests, GE crops can prevent yield losses compared with non-GE hybrids, particularly when pest infestation occurs. This effect is particularly important in the case of Bt crops. Before the commercial introduction of Bt corn in 1996, the European corn borer was only partially controlled using chemical insecticides. The economics of chemical use was not always favourable, and timely application was difficult. For these reasons, many farmers accepted yield losses rather than incur the expense of chemical pesticides to treat the insect. Consequently, the use of Bt corn often resulted in yield gains rather than pesticide savings. On the other hand, a different Bt corn trait selected for resistance against the corn rootworm, previously controlled using chemical insecticides, may provide substantial insecticide savings. This new Bt corn variety was recently introduced commercially.

An ERS study estimated the impact of adopting GE crops on yields using an adoption model and 1997 survey data. The study shows that an increase of 10 per cent in the adoption of HT cotton led to a 1.7-per cent increase in yields. Similarly, the adoption of Bt cotton in the Southeast was related to a significant increase in yields. On the other hand, the adoption of HT soybeans was related to only small (but still significant) increases in yields.

Adoption, Net Returns, and Household Income

According to an ERS study, the impacts of GE crop adoption on U.S. farmers vary by crop and technology (Fernandez-Cornejo and McBride, 2002). The main results of the ERS study are presented below:

- Planting HT cotton and corn was associated with increased producer net returns, but HT corn acreage was limited. The limited acreage on which herbicide-tolerant corn has been used is likely to be acreage with the greatest comparative advantage for this technology. The positive financial impact of adoption may also be due to seed companies' setting low premiums for herbicide-tolerant corn relative to conventional varieties in an attempt to expand market share. Limited adoption of HT corn may be due to constraints imposed on rotation with soybeans. Also, some HT corn varieties have limited approval outside the U.S., restricting their export market potential.
- Adoption of Bt cotton and corn was associated with increased returns when pest pressures were high enough. The adoption of Bt cotton had a positive association with producer net returns in 1997, but the association was negative for Bt corn in 1998. This

suggests that Bt corn may have been used on some acreage where the value of protection against the European corn borer (ECB) was lower than the premium paid for the Bt seed. Because pest infestations differ across the country, the economic benefits of Bt corn are likely to be greatest where target pest pressures are most severe. The decision to use Bt corn must be made before observing the ECB pest pressure, and damage caused by the ECB varies from year to year. Some farmers may incorrectly forecast infestation levels, corn prices, and yield losses due to infestations, resulting in "overadoption." Also, producers may be willing to pay a premium for Bt corn because it reduces the risk of significant losses if higher-than-expected pest damage does occur.

- Despite the rapid adoption of HT soybeans by U.S. farmers, no significant impact on net farm returns was evident in 1997 or 1998. This lack of profitability suggests that other factors may be driving adoption for many adopters, such as the simplicity and flexibility (less management time) of weed control. This implies more time available to off-farm employment by farm operators and their spouses. (On average, off-farm earned income is more than twice the net income earned from farming.)
- Recent ERS research using 2000 data showed that adoption of HT soybeans was associated with significantly higher off-farm household income for U.S. soybean farmers. Onfarm household income was not significantly related to adoption, but total farm household income is significantly higher for adopters.

In addition, ERS research suggests that, controlling for other factors, pesticide use declined with adoption. The overall reduction in pesticide use associated with the increased adoption of GE crops (Bt cotton; and HT corn, cotton, and soybeans, using 1997/1998 data) also resulted in a significant reduction in potential exposure to pesticides. The decline in pesticide applications was estimated to be 19.1 million acre-treatments (Fernandez-Cornejo and McBride, 2002). Total pesticides applied to corn, soybeans, and cotton declined by about 2.5 million pounds (active ingredients), despite the (slight) net increase in the amount of herbicides applied to soybeans.

Adoption and Conservation Tillage

The environmental impact of conservation tillage (including no-till, ridge-till, and mulch-till) is well documented. Conservation tillage reduces soil erosion by wind and water, increases water retention, and reduces soil degradation and water/chemical run-off.

According to USDA survey data, the portion of acreage planted with HT soybeans under conservation tillage was larger than the portion of acreage growing conventional soybeans. About 60 per cent of the area planted with HT soybeans was under conservation tillage in 1997, versus 40 per cent of conventional soybeans.

Differences in the use of no-till between adopters and nonadopters of HT soybeans are even more pronounced: 40 per cent versus 20 per cent. As a result, adoption of HT crops may indirectly benefit the environment by encouraging the adoption of soil conservation practices that control soil erosion, soil degradation, and run-off.

Economic Benefits of GE Crops

GE crops can offer producers distinct advantages over conventional varieties, such as higher yields and lower pest control costs. But producers are not the only ones to gain from the adoption of GE crops. Biotechnology developers and seed companies gain by charging technology fees and seed premiums to adopters of GE varieties.

Ultimately, U.S. and foreign consumers may benefit from GE crops through lower commodity prices, which result from increased supplies.

ERS estimated the total market benefit arising from the adoption of three biotech crops in 1997: herbicide-tolerant soybeans, insect-resistant (Bt) cotton, and herbicide-tolerant cotton. Estimated benefits were around $210 million for Bt cotton, $230 million for HT cotton, and $310 million for HT soybeans (Price *et al.*, 2003). This benefit includes the change in total welfare in both the seed input and commodity output markets. Estimated benefits and their distribution depend particularly on the analytical framework, supply and demand elasticity assumptions, crops considered, and year-specific factors (such as weather).

There are tangible benefits to farmers who adopt first-generation GE crops. Not all of the benefits are reflected in standard measures of net returns. As in all studies, results should be interpreted carefully, especially since the impact studies are based on a few years of data.

The impacts of GE crops vary with several factors, most notably annual pest infestations, seed premiums, prices of alternative pest control programs, and any premiums paid for segregated (i.e., non-GE) crops. These factors will continue to change over time as technology, marketing strategies for GE versus conventional crops, and consumer perceptions evolve.

GM Seeds

With increase in population and concern about the quality of food, the bio-agriculture has gained focus in the recent past in India. The farmers in India are looking at the GM seeds, biofertilizers, biopesticides from which they can expect more return on their investments and also increase productivity. The farmers are now getting a premium for the organic produce. With this now they are able to export their produce at much higher price. Lot of research initiatives are also being undertaken in the field to explore new technologies and new tie ups and joint ventures are being set up and are getting approvals from government for exploiting the technologies for the betterment of farmers and obtaining decent returns

The challenge of producing more food grains to feed the ever increasing population of India that has already crossed one billion mark with less resources has bought companies like Mahyco, Monsanto, Syngenta, ProAgro, Advanta to invest in GM crops. It was in 2002 a joint venture between Mahyco and Monsanto called Mahyco- Monsanto Biotech Ltd got the green signal from the Government of India for the commercial production and sale of Bt cotton (Bollgard) in six southern states of India.

A lot of awareness campaigns have to be conducted to reach out to the farmers to brief them about the benefits of using seeds that are resistant to pests, diseases, herbicides, and crops which are tolerant to drought, cold, salinity and other harsh environments. This will bring in confidence among the farmers as well as the industry people.

Biofertilizers, Biopesticides

In addition to GM seeds, the farmers are also looking at biofertilizers, biopesticides to get more benefits. Now the farmers are using formulations based on Bt, viruses like NPV, and GV, as well as neem-based pesticides.

To meet the increasing demand, the industry has to scale up investments in biofertilizers and biopesticides. Conservative estimate shows that the 10 per cent saving through the use of biofertilizers will result in an annual saving of 1.094 million tons of nitrogenous fertilizers costing around Rs 550 crore.

Biotechnology in Food and Agriculture

Biotechnology provides powerful tools for the sustainable development of agriculture, fisheries and forestry, as well as the food industry. When appropriately integrated with other technologies for the production of food, agricultural products and services, biotechnology can be of significant assistance in meeting the needs of an expanding and increasingly urbanized population in the next millennium.

There is a wide array of "biotechnologies" with different

techniques and applications. The Convention on Biological Diversity (CBD) defines biotechnology as:

> *"Any technological application that uses biological systems, living organisms, or derivatives thereof, to make or modify products or processes for specific use".*

Interpreted in this broad sense, the definition of biotechnology covers many of the tools and techniques that are commonplace in agriculture and food production. Interpreted in a narrow sense, which considers only the new DNA techniques, molecular biology and reproductive technological applications, the definition covers a range of different technologies such as gene manipulation and gene transfer, DNA typing and cloning of plants and animals.

While there is little controversy about many aspects of biotechnology and its application, genetically modified organisms (GMOs) have become the target of a very intensive and, at times, emotionally charged debate. FAO recognises that genetic engineering has the potential to help increase production and productivity in agriculture, forestry and fisheries. It could lead to higher yields on marginal lands in countries that today cannot grow enough food to feed their people. There are already examples where genetic engineering is helping to reduce the transmission of human and animal diseases through new vaccines. Rice has been genetically engineered to contain pro-vitamin A (beta carotene) and iron, which could improve the health of many low-income communities.

Other biotechnological methods have led to organisms that improve food quality and consistency, or that clean up oil spills and heavy metals in fragile ecosystems. Tissue culture has produced plants that are increasing crop yields by providing farmers with healthier planting material.

Marker-assisted selection and DNA fingerprinting allow a faster and much more targeted development of improved genotypes for all living species. They also provide new research methods which can assist in the conservation and characterisation of biodiversity. The new techniques will enable scientists to recognise and target quantitative trait loci and thus increase the efficiency of breeding for some traditionally intractable agronomic problems such as drought resistance and improved root systems.

Regulations on Biotechnology and Agriculture

While the Canadian Food Inspection Agency is the lead agency responsible for the regulation of agricultural products, it is not the only agency with this responsibility. Health Canada reviews novel products for food safety and sets data requirements for the assessment of the safety of all foods.

Health Canada also identifies hazards and specifies the standards that food inspectors observe. Its Pest Management Regulatory Agency (PMRA), assumed responsibility for registration and regulation of all pest control products in April 1995. It evaluates any product having pesticidal properties. Environment Canada works with regulatory agencies to help develop standards required for products that may affect the environment.

A wide range of products are regulated under the authority of the Canadian Food Inspection Agency (CFIA).

Examples of Products Regulated by the CFIA

- Agri-food products,
- Veterinary vaccines and biologics,
- Plants and animals,
- Fertilizers,
- Livestock feeds, and
- Seeds.

The regulatory authorities for these products are contained in Acts and Regulations. Whether the product has been produced by conventional methods or by advanced biotechnology, the general information requirements are the same. The agency strives to ensure that these regulatory requirements are consistent with those of recognised international scientific groups and with other national governments.

Guidelines for Evaluating Products

Product evaluators at the Canadian Food Inspection Agency have developed guidelines that are consistent with those used by international authorities. Some of the principles that are followed in Canada are:

- to build on current legislation where possible, rather than creating new legislation to govern new products which are developed;
- to focus on product characteristics, rather than the method of production. At the present time, all products developed through genetic engineering (recombinant products) are assessed for unintended effects that may result from the introduction of foreign genes or DNA sequences;
- to conduct evaluations for each product on the basis of its unique characteristics and to establish appropriate safety levels based on the best scientific information. Safety is defined, not as the complete absence of risk, but rather as the level of "acceptable risk". If the risk is not acceptable, the application will be refused.

Scientific information is increasing continually and regulators must build their regulatory approach on this changing information base. This results in more efficient assessment as similar products are developed over time and could also mean expanded uses and additional applications for a product as new information becomes available.

Specific Product Groups

Some of the key control procedures that apply are also listed:

1. Plants with novel traits are one of the more active areas of research using both traditional and genetic

engineering (recombinant) methods. Crop and horticultural plants are included in this group and are regulated under the *Seeds Act* and the *Plant Protection Act*. Risk assessments are conducted on plants with novel traits and consider plant biology, the new characteristics, the potential environmental impact and how the plant might affect human or animal safety. In evaluating the application, regulators may request data generated from controlled field trials.

2. Biofertilizers include rhizobia, other types of free-living nitrogen-fixing bacteria and some fungi. Recombinant products are not yet commercialised in Canada, and the research focus is on genetically improved rhizobia. The *Fertilizers Act* requires that products must be registered and specifies standards and labelling requirements.
3. Feeds are defined as any substance or mixture of substances, manufactured, sold, or represented for use for consumption by livestock, for providing the nutritional requirements of livestock, or for the purpose of preventing or correcting nutritional disorders of livestock. Novel feeds include microbial products (both viable and non-viable), plants with novel traits and fermentation products such as enzymes, biomass proteins, amino acids, vitamins and flavouring ingredients.
4. Veterinary Biologics include animal vaccines, toxins, antisera and diagnostic kits. Currently there are two classes of recombinant products: those inactivated products prepared from genetically engineered organisms; and those products containing live recombinant organisms. The *Health of Animals Act* requires extensive testing, limited field trials with target species, and ongoing quality assurance monitoring of the manufacturer. Licensing is also required. Some categories of veterinary biologics are regulated by

Health Canada under the *Food and Drugs Act,* because these are prescribed substances, such as hormones like rbST.

5. Food inspection is a broad area which covers meat, dairy products, eggs and egg products, fruits, vegetables, honey and maple products. The Canadian Food Inspection Agency provides inspection programs that enforce safety standards, review labelling, and monitor the product quality and marketing. Genetically engineered foods require a full risk assessment and Health Canada establishes safety standards and specifies labelling requirements for safety under the *Food and Drugs Act*.

6

Biotechnology and Sustainable Development

Genetically modified (GM) crops abound; knowledge and information, which were once public property, are now privately patented; and, as an extension of this, much public research has become privatised. Whilst many things have changed over these past 12 years, the need for concise, reliable, impartial and valuable information, articulating the relevance, process and impact of biotechnology on developing countries, remains the same. In fulfilling this need the Monitor continues to occupy a unique position.

As Guido Ruivenkamp points out, monitoring is needed to give people the knowledge and capability to influence processes that otherwise might lead to an everwidening gap between the industrial transformation of agriculture and the social context in which it occurs.

He notes that appropriation and substitution – two important long-term processes of industrialisation – have become important in biotechnology, leading to the industrial transformation of agriculture. The relevance of his analysis is underlined by Raymond Feddema, who was able to acquire a copy of South Korea's master plan for its future in biotechnology. After having successfully followed a path of industrialisation, by 2010 the country plans to become one of the world leaders in biotechnology.

Biotechnology is no longer an issue of the future, as it was in the 1980s, but is embedded in specific social processes. The industrial transformation of agriculture and biotechnological developments are influencing each other. The economic concentration in the seeds sector has reached an unexpectedly high level and the relationships both between farmers and seed companies and between farmers and food processing companies are changing. In the past the Monitor has described how biotechnology is interwoven with these social processes. In the near future the Monitor may also pay attention to initiatives that aim to reconstruct biotechnology to locally-specific initiatives.

Twelve years ago the Biotechnology and Development Monitor was launched to inform the public about the potential impacts of biotechnological developments on the social economic situation of farmers, especially in developing countries. In view of the analysis of the introduction of biotechnology in the agroindustrial production chain (Ruivenkamp 1984-1989), the necessity was felt to warn organisations in developing countries both of the potential trade shifts in agricultural products and of the shifts in power relations between farmers' organisations and transnational food and seed companies. The analysis has made clear that the development of biotechnology is not isolated but on the contrary—strongly interwoven with two long-term, historical processes of the industrial transformation of agriculture: appropriation and substitution. However, the analysis has also shown that biotechnology is not only influenced by the specific context in which it is developed, but that it also influences and even modifies these processes. So, it is clear that neither a social nor technological deterministic approach should be followed but an approach in which the mutually influencing relationship between context and technology is taken seriously. Concerning the industrial appropriation of some biological activities of farming, such as soil fertility and the reproduction of plants through the supply of chemical fertilizers and hybrids, the analysis has shown that biotechnology is influenced by this process as well as reinforcing and changing it.

Through the development of tissue culture, cell fusion and r-DNA techniques the Ruivenkamp argues that the *Tailor-made Biotechnologies Network* is an important means to build up local capabilities and knowledge. Johanna Ulmanen compares Dutch and Swedish approaches to ensuring local capacity building. She shows that while both approaches to development have the same goal – to improve the livelihood of people – they focus their activities on different target audiences, i.e. DGIS on farmers, SIDA on researchers. In a regional seminar organised for the Monitor, Marilyn Minderhoud Jones describes an approach that brought these two target audiences together, as well as other audiences such as journalists.

The link between policy-makers and researchers is further explored by George Essegbey, who shows how Ghana is trying to come to terms with biotechnology, and develop an appropriate policy accordingly. Rajeswari S. Raina, discussing innovation capability in India, makes a case for the involvement of social scientists, who are becoming increasingly important not only in the process of translating developments in the science-industries to a larger public, but also in policy research and development. What it means to take stakeholders on board in the field when tailor-made biotechnologies are developed is shown by Jasper Buijs. The process of building up networks, alliances and organisations is not without its problems and Frederik Oberthur makes a case for overcoming predetermined thinking.

When it comes to exploring additional tools for monitoring, several authors point to the potential of exploiting online technologies for biotechnology and development. The Community Biodiversity Development and Conservation Programme is a global network preparing itself for its next phase of activities, by using interactive technologies to build up knowledge and capabilities in the South. Robert Chakanda explains that this includes the development of an online course on biodiversity and conservation. Interactive and new technologies help develop new channels and forge new networks and alliances, and thus cut across traditional boundaries of space, time and scope, argues Dai Forterre. The Monitor would like to have a similar global impact in biotechnology.

The Monitor's unique constituency—at present 5000 subscribers in 130 countries, two thirds of whom are in developing countries—is a network of networks in its own right. The Monitor, entering a new phase in its own development, wants to use additional means to disseminate information.

Impact of Terminator Technology

Genetic use-restriction technologies enable the developers of transgenic plants or animals to protect their variety or breed from unauthorised use in a biological way. The use of 'terminator technology' can have different impacts on farmers and breeders. If the technology is effective, it impacts on agrobiodiversity and environment.

In March 1998, the US Patent Office granted the US patent 5,723,765 on the "Control of Plant Gene Expression" to *Delta and Pine Land Co.* and the *US Department of Agriculture*. The patent describes a set of interacting genetic elements that allows the controlled expression of value-added traits or of seed viability in a crop plant. Control of the expression may either rest with the owner of the plant variety or with the farmer who grows the variety.

The *Rural Advancement Foundation International* (RAFI), now called Etc. group, subsequently coined the term *terminator technology* to stress one of the likely applications of this technology which is that it becomes possible to grow crops which have seeds that are viable when sold to the farmer, but of which the seeds from the subsequent harvest are sterile. As a result, farmers would be unable to maintain a commercial variety from their own seed stocks and would be forced to return to the seed provider. In an earlier study by Richard Jefferson and co-authors for the secretariat of the *Convention on Biological Diversity* (CBD) on the consequences of the new technology, the term *Genetic Use Restriction Technologies* (GURTs) was proposed.

Genetic Use Restriction Technologies

A substantial number of patent applications now describe various GURT concepts and elements. Two types of technologies can be distinguished. On the one hand, technologies can restrict the use of a specific trait by regulating its expression (T-GURTs). In T-GURT concepts, one or more genes conferring a single trait are switched on or off at will through chemical inducers. The seed itself remains viable.

On the other hand, technologies can restrict the use of the entire variety through interference with reproduction (V-GURTs). Three types of V-GURT strategies can be distinguished.

- In the first strategy, the seeds are fertile, but through the application of a chemical a dormant 'lethal' gene can be activated. This gene inhibits full seed development. As a consequence, seeds are fit for consumption but infertile.
- In a second, alternative strategy, a lethal gene is expressed in the seed, resulting in sterility. Breeders can apply a chemical compound that activates another gene to safeguard the fertility of the seed and the reproduction of the variety, but they will stop doing so before selling the seed.
- The third, and complementary, strategy concerns vegetatively reproducing crops, e.g. root and tuber crops and ornamentals. The described method involves

prevention of growth during storage, which may be in the interest of breeders, growers and/or consumers. Normally, a gene that blocks growth is expressed, but growth can be restored by activation of a second gene. Regulation of hormone metabolism and function lies at the basis of this strategy.

It is important to realise that, although current patent applications apply to plants, GURTs can be built into any organism, including farm animals, fish and forest trees.

Other methods that share some degree of 'use restriction' with V-GURTs are hybrid technology, but also triploidisation and the introduction of male or female sterility. Differences between these latter technologies and V-GURTs are that the germplasm remains available to farmers and competing breeders for further breeding. In other words, if V-GURTs can be made sufficiently effective, they largely rule out the Breeder's Exemption and the Farmer's Privilege on the use of protected varieties as described in Plant Breeder's Rights.

Motives for GURTs

The application of GURTs can be inspired by different, not mutually exclusive motives. These motives fall in the following categories:

- *Property Protection Interests*: Breeding companies wish to safeguard their investments in improved varieties, whether produced by classical breeding or genetic engineering. GURTs may offer a better biological insurance against the free use of genetic innovations than patents, plant breeder's rights or licences. Hybrid seed, stemming from two genetically different paternal and maternal lines, offer a certain protection of germplasm against reproduction by farmers, because the seeds normally result in different and heterogeneous plants in the next generation. For cross-fertilizing crops, this prevents straightforward regeneration of elite varieties. For many self-fertilizing crops, hybrids are not feasible or effective. Therefore, these crops represent the primary targets for the use of V-GURTs in order to obtain biological property protection. Crops concerned include rice, wheat, soybean and cotton, and horticultural crops as well as ornamentals that are vegetatively multiplied. T-GURTs to protect high-value added traits in newly released commercial varieties may be applied to virtually all crops.

- *Environmental Interests:* GURTs can be used for the environmental containment of entire transgenic varieties (V-GURT) or of specific transgenes contained in transgenic varieties (T-GURT). In both cases, the focus will be most likely on species that may establish themselves in ecological niches, and/or for which wild relatives exist in involved crop production systems.

T-GURTs can be used to contain traits with suspected health risks for the farmer or consumer, or traits that may somehow form a threat to biodiversity or other aspects of the environment. An example is a pathogen resistance gene that may outcross to weedy relatives and thus increase the fitness of those weedy relatives.

- *Production interests*: It can be in the producer or farmer's interest to restrict the expression of a trait for example to a specific phase in the development of the plants or animals or during biotic or abiotic stress. Stress response takes a lot of resources and may diminish the quality of the processed product and is only desirable when really needed. Alternatively, V-GURTs may be used to control successful reproduction of farm animals in order to safeguard the integrity of adapted crossbreeds, resulting from the mating between local and high-yielding commercial breeds.

It is important to distinguish between the impacts of GURTs on farmers vis-a-vis those on breeders. It is more likely that access to technology circumventing GURT protection will be more available to breeders than to farmers. Farmers will have fewer options to choose from.

Will GURT Application be Successful?

The increasing pace at which genetic information from genomics research is becoming available will most likely result in several effective GURT prototypes within five to ten years. Still, several bottlenecks in current concepts will prevent the application of GURTs in a number of cases. A major one is the requirement for watertight regulation of the genes. For instance, if a lethal gene is not only expressed after induction but also continuously at a low level, this may decrease seed viability.

A number of technical factors will determine for what purposes, on which crops, trees, farm animal and fish breeds, and on what time scale GURTs will be developed and applied. Costs to solve technical bottlenecks as well as costs stemming from regulatory requirements will be juxtaposed against expected rates of return when decisions are made on investments in GURT development. Application of GURT in crop breeding can be expected to precede its applications in the breeding of trees, fish and farm animals, as is also reflected in current patent applications.

Options for introduction of GURTs in tree species are technically similar to those in crops. Applications in aquatic species may also emerge soon, although economic and environmental issues, such as the threat to survival of local indigenous fish populations with which the transgenic fish can interbreed, may result in a slower or more restricted development. The application of GURTs in animals currently experiences much higher technological as well as ethical barriers. Currently, breeding efforts focus on crops and animal breeds for which hybrid technology, triploidisation, male or female sterility and/or the legal provisions of *intellectual property rights* (IPRs) are available. A major influence of GURT availability might be a shift towards crops and animals with the largest market shares or highest profit margins since the appropriation of breeding investments by the use of GURT will be independent of availability of those other technologies.

Furthermore, it can be envisaged that T-GURTs will prove to be more widely acceptable and technologically less demanding than V-GURTs because control may be less absolute.

Stakeholders and Impacts on Agrobiodiversity

The way in which the conservation of agrobiodiversity has been 'organised' has major consequences for assessing the potential impact of GURTs on agrobiodiversity and agroproduction systems and justifies a stakeholder analysis.

Stakeholders encompass the public and private sector, as well as farmers and supportive *non-governmental organisations* (NGOs). Obviously, each of these stakeholders has different roles and interests. For example, the private sector largely maintains its own *ex situ* collections of germplasm and other commodities or has ready access to such stocks, while autonomous farming systems only marginally rely on external sources of germplasm and other commodities and largely maintain their own knowledge system.

Stakeholders in the private sector do not generally consider it to be a priority to maintain agrobiodiversity in autonomous farming systems, or to ensure the survival of wild relatives in natural ecosystems as a much wider source of genetic resources. Although the private sector may regard genetic diversity in such farming systems as highly relevant in the long run, it is often regarded as being mainly a public, long-term responsibility.

With increasing GURT application, access for autonomous farming systems to novel industrial innovations under GURT control will become limited or even absent. The degree to which this phenomenon will affect these farming systems will be dependent on their current use of such new traits and varieties. In other words, the issue at stake here is the genetic overlap: which crops and animal breeds form major constituents of both industrial and autonomous farming systems? To what extent does industrial and autonomous production under similar agroecological conditions result in breeding investments in varieties and breeds used by both sectors? Which of the traits developed by the private sector are relevant to small-scale producers?

At first sight, interference between GURT development and the maintenance of agrobiodiversity may seem limited. However, on a larger time-scale effects may be substantial. Farmers in autonomous production systems mainly rely on their own seeds and animals adapted to their particular agroecosystems, their farming practices and cultural preferences. This does not mean that there is no utilisation of varieties and breeds stemming from the private sector.

On the contrary, as far as accessible, these varieties and breeds are used by farmers to test them under their own conditions and—if suitable—to recombine them with their own germplasm to improve and 'breed' new varieties and breeds. In addition, the public sector, the *national agricultural research systems* (NARS) and the institutes of the *Consultative Group on International Agricultural Research* (CGIAR), act as a channel by which major genetic improvements developed in the private sector can reach autonomous farming systems. In conclusion, crop development in farmers' seed systems may miss options for crop development available to industrial seed systems due to limited access to new traits and varieties protected by GURT.

Environmental Effects

Negative effects on yields of outcrossing from V-GURT containing varieties to neighbouring crop stands may occur. The size of these effects will depend on the outcrossing rate of the crop concerned, and on the physical distance between donor and acceptor plants, related to the intensity of the farming systems.

These negative effects will also occur in those cases in which VGURT strategies would be consciously employed to prevent unwanted gene flow from *genetically modified organisms* (GMOs) into the environment, cultivated or wild, as an approach to avoid undesirable effects of a deliberate release of GMOs. Such negative effects are most likely to occur in outcrossing crops and trees. The probability might be the highest for trees because of the ability of pollen to disperse over long distances. However, negative impact may be low assuming that tree seeds do not form an economically important product.

Potential negative environmental effects in animal husbandry can be avoided more easily, in view of the high level of domestication and current farm practices to control reproduction. In contrast, the transfer of GURT constructs into wild fish populations with potential negative effects on population survival might be substantial because of the low level of domestication and the high probability of escapes of fish varieties.

For T-GURTS, the impact of outcrossed constructs will be limited in most cases. It can be expected that most GURT-protected traits will be under the positive control of a chemical compound, a so-called inducer. If such a construct would outcross to other crops to which the chemical is not applied, the construct will usually remain unnoticed. However, such constructs may still survive and exceptions to this rule can be imagined. *First*, the trait could be induced by a range of related compounds or triggered by naturally occurring events, such as steroids, pest and disease infestations. *Second*, the receiving organisms could already contain a GURT construct controlled by related inducers so that the inducer would activate both constructs. Such undesirable effects might be avoided by choosing highly specific inducers.

A *third* and more problematic scenario is the outcrossing of GURT constructs that exhibit negative control of the involved trait. This means that a trait is expressed unless it is blocked by the application of an inducer. In the case of outcrossing, the chemical compound would most likely not be applied to the organism that newly integrated the construct and the trait would be continuously suppressed, also when not desired. Such outcrossing could affect not only domesticates but also wild relatives.

Although individual elements of a GURT construct may outcross independently and exhibit certain effects, the likelihood of such events is very low, since these assume both a recombination event and an outcrossing event in most cases. In other words, such elements will only outcross separately from the other elements of the same construct, once they have been physically separated from each other in the genome.

To safeguard long-term on-farm conservation and development of plant genetic resources, increased investments in public plant breeding, including *participatory plant breeding* may be needed to correct the increasing gap in absorption of innovations by autonomous farming systems. This is most important for all staple crops and major vegetables and fruits grown in industrial and autonomous farming systems. Similar assumptions can be made for applications of GURTs in the other agricultural sectors. In particular in aquaculture, some species such as tilapia and crustaceans are grown in both small-scale and large-scale systems. The growing dependence of the small-scale sector on seed developed and provided by industry may have negative effects on the biodiversity of species maintained in aquaculture. Alternatively, the small-scale sector might have to rely completely on its own germplasm improvement, thereby restricting genetic resources development. Applications of V-GURTs in crops, whether to protect investments or to protect the environment should be assessed for their negative yield effects. Also, the probability of negative effects on local fish populations and tree relatives of transfer (introgression) of V-GURT constructs warrant further study. The impacts of negatively regulated traits with undesirable effects on yields and quality in a wide sense may need further discussion and policy development.

The use of some compounds as inducers, for example steroids, may have to be regulated, like pesticides and veterinary medicines, depending on the nature of the compounds. Their effects on the target organisms as well as the environment and human appliers and consumers need to be assessed.

Bt Cotton

The development of agricultural biotechnology offers new opportunities to increase agricultural productivity. Some of the potential gains that can be achieved with this technology in the developed countries include higher yields, enhanced product quality and lower pesticide and labour costs. Such gains are particular important in developing countries, especially in the fight to alleviate poverty.

So far there has been little evidence as to whether agricultural biotechnology can benefit farmers in developing countries. The aim of the study outlined here was to provide some early evidence on the economic implications of adopting a genetically modified crop, Bt cotton, at farm level in a developing country—the Makhathini Flats region of the Republic of South Africa. The study concentrated on the uptake of the technology by smallholder farmers and on the economic gains, if any, that they derived from adopting Bt cotton.

The worldwide area under commercially grown *genetically modified* (GM) crops has been rapidly increasing since they were first introduced in 1996. Insecticide and herbicide resistant traits account for more than 60 per cent of the GM crops grown worldwide. In South Africa, the commercial release of insect resistant Bt cotton and Bt maize was made possible by the *Genetic Modified Organism Act* (GMO Act 15) of 1997. Bt cotton was the first commercial release of a GM crop variety in sub-Saharan Africa.

In 1999/2000 a total area of 100,000 hectare of Bt cotton was grown in South Africa by 1530 commercial farmers and 3000 small-scale farmers mostly under dryland conditions in the Northern Province, with some in KwaZulu-Natal and the Free State. Since 1998, smallholder farmers in the Makhathini Flats (KwaZulu-Natal province) have been adopting the genetically modified cottonseed variety *NuCOTN 37-B* with *Bollgard*. Sixty per cent of these farmers have plots of between 10 and 20 hectare. Seed companies estimate that in the 2000/2001 growing season, around 95 per cent of the 4000 smallholder farmers in the Makhathini region will have adopted this Bt cotton variety. Agriculture is an important source of income in the Makhathini area, and rural households typically occupy between 1 and 3 hectare of land allocated to them by their tribal chiefs. The major crops in the area are beans, maize and cotton. The latter usually occupies most of the farm and is grown as a commercial crop.

Planting takes place from mid-October to mid-December and harvesting from mid-May to mid-June. One of the main reasons for growing cotton is that the crop needs less intensive management than maize or beans and can survive fluctuating weather conditions. The main constraints on market-orientated agriculture in the region are a labour force depleted because of male migration to towns, low capitalisation because of a heavy reliance on credit, the climatic constraint of erratic rainfall, and pest attack.

Vunisa Cotton is at the heart of the farming system in Makhathini. Vunisa Cotton is a private, commercial organisation supplying seed, fertilizer, pesticide, credit and information to smallholder farmers in the region. It also then buys their cotton after harvest. Seed companies such as *Delta Pineland, Clark Cotton* and *OTK*, and agrochemical companies such as *VenChem* supply their products to Vunisa Cotton, which then retails them to the small farmers.

In the case of Bt cotton, Monsanto owns the Bt gene that Delta Pineland has used to develop the NuCOTN 37-B with Bollgard variety. The Landbank of South Africa provides the finance (credit) and Vunisa Cotton is responsible for allocating finance to farmers following stringent assessment that takes into account their assets, personal finance, experience, repayment history, and yield assessment over time. Information on cotton is disseminated to the farmers via extension personnel employed by Vunisa Cotton. Each of the extension staff is responsible for a specific area and regularly visits the farmers and holds meetings.

Although there is a government extensionist next door to Vunisa, they are not responsible for cotton production. This is left the hands of the Vunisa. There has been no research done so far, as they are private extensionists. All the farmers deliver their cotton to Vunisa Cotton for weighing and grading before payment is made. Vunisa Cotton is the sole supplier of cotton inputs and buyer of cotton output in the region in the northeast of KwaZulu Natal. As such, it plays a pivotal and vital role in the farming system. Vunisa, however, does not provide inputs, credit and marketing for other crops grown in Makhathini.

Survey Data

The study was conducted in November 2000. It is based on a questionnaire survey of 100 smallholder farmers in the Makhathini Flats. A stratified sample was taken of smallholder farmers according to whether they had adopted the Bt cotton variety. The questionnaire was completed during face-to-face interviews in the field and at Vunisa Cotton premises. The questionnaire was designed to obtain information on the physical characteristics of the farm; characteristics that defined the farmer; cropping patterns; input use and costs; cotton output and revenue; and other income sources and assets.

The aim of the study was to examine the factors involved in the adoption of Bt cotton as well as its impact on yields, gross margins and technical efficiency—notably pesticide use. Our analysis compares Bt cotton farmers (adopters) with non-Bt cotton farmers (non-adopters). The survey covered two growing seasons: 1998/1999 (first year) and 1999/2000 (second year). In the 1998/1999 season, only 10 per cent of the 4000 farmers in the region had adopted the new Bt cotton variety, while by 1999/2000 this had risen to 40 per cent.

The average farm size of respondents was 6 ha, although 62 per cent of the farms included in the survey were less than 5 hectare. Approximately 73 per cent of 100 respondents owned livestock such as hens, sheep, goats and 25 per cent had non-farm income sources. The average is 6 hectare for the sample, and most farmers own between 0.5 to 3 hectare. This is very typical for the Makhathini region where land is allocated to farmers by the chief. No one actually owns the land, but the chief has control over it. If a farmers does not use the land allocated to him, someone else can to do so. They farm small areas of cotton because of the lack of labour and credit, and cotton is only a cash crop. Some migrants work in nearby towns and most of them move to Johannesburg or Durban in search of work.

Farmers faced several problems. Amongst the agronomic problems, 57 hectare of farmers consider pests to be their biggest problem and 62 hectare of these categorised the bollworm as the major pest. Too much rain was a more distant second at 24 hectare, followed by weeds, which was ranked highest by only 11 per cent. Among the non-agronomic problems, 82 hectare of farmers cited lack of credit as their major non-agronomic constraint, as compared with 14 hectare who rated land scarcity as their biggest problem and only 4 hectare whose greatest difficulty was lack of labour.

Adoption of Bt Cotton

The stratified sample consists of 19 adopters and 81 non-adopters in the first year and 60 adopters and 40 non-adopters in the second year. All 19 farmers who grew Bt cotton in the first year carried on growing it in the following year, suggesting that they were satisfied with the variety. Generally, the adopters of Bt cotton were more experienced and had larger farms. Soil differences are negligible; the whole area has similar soil and used to be a flat riverbed before the construction of the huge Jozini Dam in 1974. None of the 4000 smallholders use irrigation in their farming.

They all plough using tractors and carry out planting, weeding and harvesting manually. There is a variation between farmers in the amount of labour and seed used. The larger farmers use more hired labour and are probably more efficient in using resources having spent more years in farming than the non-adopters. A combination of these two characteristics ensured that these farmers were specifically targeted by Vunisa Cotton agents for the promotion of Bt cotton in 1998/99—the first year of release—and were therefore more likely to be exposed to the technology and receive credit. The key factors affecting early adoption of Bt cotton were thus the availability of credit, or other means of purchasing inputs, such as non-farm income, and pressure from the Vunisa Cotton personnel.

When questioned in 2000, after one season of experience, about why they adopted Bt cotton or might consider adopting it in the future, 44 per cent of respondents directly cited savings on the cost of insecticide as the main reason, while 24 per cent cited expected increases in yield. Approximately 10 per cent said that the expectation of less time spent spraying Bt cotton was critical in their decision to adopt it.

Most of the farmers surveyed did not identify any problems with Bt cotton. Until recently they were not given any training on how to grow it. However, now information is being made available, for example, on the subject of refuges. Almost 90 per cent of the non-adopters in 1999/2000 were willing to adopt the technology in the future, but cited cost of the seed as the main reason for not adopting it at the moment. As already noted the vast majority of respondents felt themselves to be constrained by a lack of capital, and Vunisa officials targeted the larger, more creditworthy farmers, hence it is perhaps not surprising that credit has played such a major role in the adoption of Bt cotton in Makhathini. All cotton is for domestic use. South Africa is not self-sufficient in cotton and still relies on imports.

Economic Impact of Bt Cotton

Bt cotton gave higher yields per hectare than other non-Bt varieties. However, there was a marked seasonal effect. In 1998/99 the average yield increase of Bt over non-Bt was nearly 18 per cent, and in the 1999/2000 season this rose to 60 per cent. These increases were despite the fact that the seeding rate of Bt adopters was only 45 per cent of the recommended rate (the equivalent rate for non-adopters was 55 per cent of the recommended rate).

Thus, Bt cotton gave even higher yields per kg of seed planted than the non-Bt crop. The shift in yield differential in the two seasons was probably related to rainfall, which can affect the incidence of bollworm or growth of the crop or even the planting time. The 1999/2000 season was a bad one due to unusually heavy rainfall, and average cotton yields fell. However, the Bt adopters suffered a fall in yields between the two seasons (of 18 per cent) much less than those who did not adopt (40 per cent). The rain could have affected the incidence of disease and pest such as bollworm and the length of the growing season. It meant a delay in planting and also affected the germination rate.

The use of Bt cotton increased seed cost per hectare by over 100 per cent in both seasons. The increase in price to Rand 464 per bag includes the technology fee of Rand 240. The price is fixed and does not vary with demand or supply. This high seed cost was only partially offset by the fact that Bt cotton reduced pesticide costs for Bt adopters in both growing seasons. On average, in the first season pesticide costs for Bt adopters were reduced by nearly 13 per cent, while in the second season the reduction was 38 per cent. It could be that in the first season some producers continued to spray either as a risk reduction strategy in case the Bt variety proved not to be as resistant as claimed, or because they received little information on the nature of the Bt variety and its pesticide attributes.

Fifty-four per cent of heads of farm household in this area are male. However, there seems to be little difference in terms of income generated between male and female headed households.

As a result of increases in yield and value of output, and reduced pesticide costs (despite seed cost increases), the average gross margin per hectare (value of output minus the cost of seed and pesticide) for the Bt crop was higher than for the non-Bt. In the first season, the average gross margin for adopters was 11 per cent higher than that of non-adopters. In the second season, adopters had an average gross margin 77 per cent higher than that of non-adopters. Input cost differential was the same for both seasons, but the main factor at play in this increase in gross margin differential was the change in yield. The second season saw a much greater yield advantage for Bt adopters relative to non-adopters.

Not surprisingly, the smaller holdings grew more intensively (with higher seed and pesticide costs per hectare) and achieved higher per hectare yields on average than the larger farms. Indeed, in the first year of adoption some producers with plantings of greater than 5 hectare of Bt cotton suffered yield and gross margin reductions compared to those growing conventional cotton varieties, while the smaller producers growing Bt cotton have substantial increases both in yields and gross margins.

By the second year of adoption, all holding sizes appear to benefit from the Bt variety, although the greatest benefit, in terms of gross margin per hectare, still accrues to the smaller producers. Judging by the data we collected there appears to be no widening inequality between Bt and non-Bt farmers. The increase in yield if sustained would not make a difference if the price of cotton were to fall. We have not as yet considered the effect of Bt on other crops. This is an early study; we are now in the process of expanding the work, finance permitting.

Despite the variation between the two seasons, the results of this survey provide considerable cause for cautious optimism regarding the impacts of Bt cotton at farm level. Although it was not so apparent in the first year, by the second year the adopters were clearly gaining economically in terms of increased yield, lower insecticide costs and hence a higher gross margin. This perhaps suggests that the advantages of Bt cotton become particularly apparent in times of environmental stress, since the second year was very wet and such conditions favour the bollworm. For risk-prone, resource-poor farmers this is an important consideration. Moreover, although our analyses are based on small sample sizes, it would appear that the smaller holdings could benefit most from growing the Bt variety. This may be a particularly important element of the technology if the aim is to reduce poverty among small farmers.

There may be additional benefits from Bt cotton that our study has not addressed. First, the analysis of savings in pesticide costs from adopting the Bt variety do not include labour costs or time saved from reduced spraying, which would be an additional benefit unaccounted for in our analysis. Moreover, a reduction in the quantity of insecticide applied would be beneficial in terms of the health of farmers and others in the area (there are numerous cases of poisonings due to pesticides) and the environment (pesticides often find their way into water courses).

In a system such as that of cotton production on the Makhathini Flats, with only one supplier of inputs and credit which is at the same time the only market for cotton, the main dynamics at play in adoption of Bt cotton are perhaps not surprising. The tendency was for the older, more experienced farmers and those with larger farms to have higher percentages of adopters especially in the first year of adoption. These farmers were particularly targeted by Vunisa Cotton staff and made aware of the Bt cotton and its presumed advantages over other varieties. They were also more likely to be granted credit or be able to finance the higher seed costs from savings or other income sources. Almost all smallholders in the sample said they would adopt Bt cotton if they had the financial resources to do so and if they had been more aware of its availability and advantages in the first season.

However, given the short time period of our study (two years), no definitive conclusions can be drawn about the adoption dynamics in the region. Some farmers may decide to return to non-Bt varieties in the long run, if the seed suppliers decide to appropriate a greater share of the benefits by raising their seed prices. If this were to happen, it would be unfortunate, as this study shows that there are potential gains at the farm level. Finally, further years of data are required for a larger sample of farmers, together with more detailed data on the labour and other aspects of adoption before final judgement of the benefits of Bt cotton to smallholders is possible. Moreover, it would be interesting to gauge the relative benefits of uptake of the crop for larger, commercial farmers compared to smallholders. This research is currently being undertaken.

Biotechnology in Developing Countries

The current global system governing biotechnology has largely been shaped by Western hopes and fears and more research is needed into how developing countries view this new technology.

In industrialised countries, psychologists, sociologists and political scientists have been researching public perception and the political dimensions of the agricultural biotechnology debate. However, few surveys have focused on how agricultural biotechnology is discussed and perceived in developing countries. This lack of research is one of the reasons why certain stereotypes have emerged in the North about how agricultural biotechnology is seen in the South. Such preconceptions make it possible for extreme proponents and opponents of agricultural biotechnology in industrialised countries to select representatives from developing countries whose views and interests coincide with their own, and encourage them to speak in the name of all the local stakeholders in their country. As a result there has often been an oversimplification of the way the risks and benefits of agricultural biotechnology are seen in the South.

Stakeholder Surveys

Between 1997 and 2001, three public attitude studies on perceptions of agricultural biotechnology in developing countries were planned and carried out in the Philippines, Mexico and South Africa. This article deals with the methodology used in these studies and discusses initial results from the Mexico and Philippines surveys. Material collected in South Africa is still being analysed. The first study focused on the Philippines. It was initiated by the *Swiss Federal Institute of Technology Zurich* (ETH) as part of a larger research project investigating public acceptance of genetically engineered food in industrialised and developing countries.

The Philippines was chosen because it had been at the centre of a controversy surrounding *Bacillus thuringiensis* (Bt) rice, a pest resistant variety developed by ETH and the *International Rice Research Institute* (IRRI). Subsequently the Swiss National Science Foundation (SNSF) approved funds for additional case studies with similar objectives to be carried out in Mexico and South Africa. These countries were selected because they had substantial local biotechnology research capacity and were the scene of considerable controversy and debate about the role of transgenic crops in local agriculture. They also each had enough regional political and economic power to influence public attitudes in neighbouring countries.

Assessing Perceptions

An individual's perceptions of the risks and benefits of a new technology is determined by personally selected sources of information, values, interests, and individual experiences. However, where agricultural biotechnology is concerned, most people cannot rely on their own experience. Instead, they depend on information received from other sources such as rumours, the experiences of people working in the field, and statements issued by industry, government, public interest groups, academia and the media. An individual's personal worldview strongly determines how information is assessed for reliability.

In each of the three case studies, a stakeholder-based approach was used to investigate public perception. The stakeholders selected were political actors who were generally recognised as playing a significant role in national agricultural biotechnology debates. It was assumed that such people would be well informed and would, therefore, have considerable influence on public opinion. This approach made it possible to conduct a survey on public attitudes in spite of a low level of public awareness of the subject. It also made it possible to analyse the way perceptions, interests and scientific knowledge shape political decisions and provided a basis for predicting how the debate might develop over time. Although it can be assumed that political actors have a significant influence in countries where little is known about agricultural biotechnology, it is important to realise that sometimes they may be ill-informed themselves. In our surveys, contradictory answers and the number of 'don't know' replies reflected the extent to which political actors were informed about the risks and benefits of agricultural biotechnology.

Selecting Political Actors

Political actors were selected with the help of key informants familiar with both the debate and its stakeholders. Moreover, member lists of national committees concerned with agricultural biotechnology and media coverage of the debate were screened in search of additional important stakeholder representatives. Most of those eventually selected to take part in the surveys were either from *non-governmental organisations* (NGOs) concerned with rural development and environmental issues and farmers' organisations, or government departments responsible for health, the environment, trade, agriculture, and science and technology. Other respondents included agronomists, biotechnologists, sociologists, and environmental scientists from local universities, representatives from agribusiness, organic agriculture and the food industry, the mass media and international organisations.

The Questionnaire

Stakeholder representatives were asked to fill in a four part, semi-standardised questionnaire. They were first asked about their perceptions of the problems facing agriculture and the potential of *genetic engineering* (GE) to solve them. Secondly, they were required to react to positive and negative statements on the risks and benefits of agricultural biotechnology. Then an attempt was made to assess how much trust they had in existing institutions and in new approaches to risk reduction and legislation.

The final section of the questionnaire included a policy network table of about 70 stakeholder organisations. Respondents were asked to assess each other on attitudes to agricultural biotechnology and their influence on public opinion, political decisionmaking and the biotechnology debate. They were also asked whether they strongly associated certain organisations with particular individuals and whether they cooperated with these organisations on matters such as information exchange and funding.

Local partners in all three countries were closely involved in drafting the questionnaires and subsequent questions clearly reflected the character of the regional debate. In the Philippines, for example, the commercial cultivation and retailing of either locally developed or imported transgenic products had been prohibited and there was a heated debate over the risks and benefits of Bt rice. The Mexican survey, however, was drafted against a background of field tests, pilot projects with both locally developed and imported *genetically modified* (GM) crops, and the discussion on introducing specific and mandatory labelling legislation to deal with imports of genetically modified Bt corn from the United States.

Reactions and Participation

A survey on public perceptions of agricultural biotechnology can easily become a political event in itself. In all three countries many political actors were suspicious of such research. Government officials were concerned about its political impact, anti-GE groups suspected hidden agendas, and scientists in private and public agricultural biotechnology research were not convinced of its usefulness. There was also some nationalist sentiment in the criticisms of those who objected to foreign research institutes initiating studies of local perceptions.

In spite of these objections and the length and complexity of the questionnaire, participation rates were surprisingly high: 65 political actors in the Philippines, 51 in Mexico and 48 in South Africa. Six months were allocated to the Philippines survey which may account for a higher participation rate there than in South Africa and Mexico where only one month was allocated because of funding constraints.

On average, 60 to 70 per cent of questionnaires were returned in each country. Our successful collaboration with local research institutes, considered trustworthy by both sides and which were able to allay nationalist objections, contributed significantly to the high participation rate. It also emerged that stakeholder representatives were motivated to take part because they were interested to find out how other stakeholders had assessed them and whether their perceptions matched average perceptions. By persistently following up questionnaires we were able to get even reluctant representatives to participate. It is significant that those who refused to take part did so not because they were suspicious of the project but because the questionnaire took so long to complete.

Data Analysis

Data analysis consisted of a descriptive analysis of prevailing perceptions, a cluster analysis presenting an evaluation of different perception patterns, and a visual representation of the principal component analysis, which portrayed the perceptions of every single stakeholder on a two-dimensional scale. The policy network table data was subject to a simplified policy network analysis that revealed the influential stakeholders in the debate and the different types of cooperation among them. After the results had been collated, a workshop was held to present the findings, discuss the implications of the results and to evaluate the appropriateness of the survey.

Initial Findings

Comparison of results to the questions put to respondents

in the Philippines and Mexico exposed some interesting differences between the two countries.

Most respondents in Mexico and the Philippines consider biotechnology to be 'just a new tool' with the potential to solve problems that currently cannot be addressed with conventional methods and to eventually contribute to future food security. While respondents in both countries disagree with the statement that GM foods pose a health risk, they expect GM crops to have a negative impact on biological diversity and are concerned that the implementation of biosafety guidelines will not be enforced properly.

Agronomic Problems

Respondents in Mexico stressed the usefulness of biotechnology in combating agronomic problems such as drought, pests and plant disease that hinder agricultural productivity.

Unlike respondents in the Philippines, they were more optimistic about the sustainability of GM crops and more sceptical of organic farming being a better alternative for resource-poor farmers. The efforts of farmers and NGOs in the Philippines to develop organic farming methods to ensure sustainable food security may be an important reason for the great expectations regarding organic farming.

Institutional Factors

In both Mexico and the Philippines there was concern about the corporate control of agricultural biotechnology. Personal interviews in the Philippines and attitudes towards statements regarding *Intellectual Property Rights* (IRPs) in the Mexican survey indicated that some stakeholders in these countries would prefer to create their own domestic research capacities and design varieties that were more appropriate to addressing problems in domestic agriculture rather than simply importing GM crops from industrialised countries.

Both Mexico and the Philippines are home to large

international agricultural research centres, such as the *International Maize and Wheat Improvement Centre* (CIMMYT) in Mexico and the *International Rice Research Institute* (IRRI) in the Philippines, which are also exploring the possibilities of GE for the genetic improvement of food crops. As public perception and acceptance of GM crops is closely related to trust, an interesting outcome of the survey was that stakeholders in Mexico tend to consider CIMMYT a trustworthy but not very influential stakeholder in the national debate whereas this was not the case with IRRI in the Philippines. IRRI is considered to be a central actor in the Philippines and its presence in the country is considered more controversial. This difference in the perception of foreign institutions has also influenced the public debates in the countries themselves.

Biosafety and Risk

Both countries were concerned about their countries' rich biological diversity. Mexican stakeholders were particularly worried about the potential of GM maize to outcross with local indigenous maize varieties. Filipino respondents expressed concern about the effects of pest resistant rice on non-target organisms. An analysis of the Mexican survey indicated that political stakeholders seemed to be informed about the discussion on the potential health and environmental risks of agricultural biotechnology but only a few were familiar with new approaches to reduce these risks.

Respondents in both countries had doubts that their national biosafety guidelines can be implemented effectively. In Mexico, respondents considered existing biosafety regulations to be inadequate whereas in the Philippines stakeholders regarded regulatory provisions to be sufficient. This difference may be explained by the fact that in 1991 the Philippines had already designed stringent biosafety guidelines for *genetically modified organisms* (GMOs). Mexico's regulatory biosafety framework mainly follows the USA though a recent initiative for a law on biosafety is currently under consideration, which includes the precautionary principle stated in the *Cartagena Protocol on Biosafety* (CPB).

Polarisation

Most respondents included in the Mexican survey were scored moderate on a scale of perceptions and, generally speaking, positions adopted were less hard than in the Philippines. Filipinos were more frequently found on the outer reaches of the perception scale indicating a stronger degree of polarisation.

In both countries, there seems to be a significant perception gap between government departments. Agencies affiliated to the Department of Environment tend to be more critical of agricultural biotechnology than those dealing with trade, agriculture and science and technology. Interestingly, it would appear that the national debate in the Philippines was already polarised before concerns arose in Europe in the late 1990s.

Stakeholder Perception

A cluster analysis of the data collected reflected the perceptions of various stakeholder groups. In the Philippines, for example, three clear patterns of perception emerged. NGOs and other public interest groups appeared to be largely opposed to GE and did not think it had any potential to solve agricultural problems. A second group made up of scientists from private and public research institutes were mainly positive towards GE and saw its potential in targeting problems caused by pests, viruses and exposure to stress. In the third cluster were government officials and politicians. They saw biotechnology as having considerable potential to solve many different agricultural problems but had an ambiguous attitude towards the potential risks and benefits of GE.

In the Mexican survey, the first cluster contained not only NGOs and other public interest groups but also representatives of government agencies (concerned with biosafety and environmental issues) and some academics. This group tended to have a negative attitude towards biotechnology. The second and largest group included scientists, government officials, politicians, NGO representatives and members of the business community and representatives of international organisations. These individuals tended to favour agricultural biotechnology and most believed in the potential of GE to solve agronomic and post-harvest problems. The third group was more strongly in favour of agricultural biotechnology and consisted of respondents from all institutional groups.

Media and Information

The policy network analysis indicated that Filipinos felt NGOs and the mass media exerted a strong influence on public opinion. Analysis of network cooperation revealed that NGOs in the Philippines play a crucial role in the acquisition and dissemination of information. Information sourced from national universities is sent to the legislature and mass media in the form of position papers. Civil society groups use public pressure created by media campaigns to influence political decisions.

This was not the case to the same extent in Mexico where public opinion, although influenced by a wide spectrum of stakeholders from different institutional settings, seemed to carry less weight. Mexican NGOs and the mass media were, however, seen as being important stakeholders in the political decision-making process.

Academia

In both countries, academia emerged as the central stakeholder in the agricultural biotechnology debate. In Mexico and the Philippines academic institutions are the most important distributors of information. They are significant players in the public debate on agricultural biotechnology and enjoy public confidence, playing a crucial role in facilitating constructive dialogues between the more antagonistic stakeholders.

Although a majority of respondents from academia have a favourable attitude towards agricultural biotechnology, they can be found in all three perception clusters. The reason why some academics have a critical attitude towards agricultural biotechnology may be that they see GE as an imported Western technology and not as the product of domestic research and development. This is particularly so in the Philippines where many local scientists are more involved with international research centres and multinational corporations than with national academia.

In addition, many academics from disciplines such as social science and ecology, work closely with Filipino NGOs in the research and development of alternative rice breeding and pest management programmes. The NGO research approach is considered to be home-grown and appeals to nationalist feelings, a factor that often plays a major role in the Filipino debate.

Feedback from Workshops

Workshops were held with respondents to evaluate the survey process and its results. A major criticism raised in both Mexico and the Philippines was that the perception of the farmers who ultimately have to choose whether or not to adopt transgenic crops had not been taken into account. This is a justifiable objection, but it was argued that these surveys were not intended to investigate the potential adoption rates of transgenic crops. The surveys sought, rather, to reach farmers through those who claimed to represent their interests in the national public debate. Whether these actors adequately represent farmers' perceptions and interests is a different question.

Stakeholder analysis indicated the political influence of the different farmer organisations and how much access they had to the political decision-making processes. It often turns out that resource-poor farmers are badly organised and under-represented in the national public debates. Many big national farmer organisations and NGOs claim to speak for marginal farmers but do not necessarily represent their interests well.

Workshop participants also stressed the importance of taking into account the opinions of all consumers and producers on transgenic food. However, the insistence that each opinion is equal in reason and value makes it impossible to find a common long-term policy that will minimise the risks and maximise the benefits of agricultural biotechnology by striking a compromise between different interests and perceptions. The failure to find a middle ground and design a pragmatic strategy may increase the real risks related to biotechnology, because political paralysis as an answer to uncertainty only creates more uncertainty.

In Mexico, several respondents criticised the selection of political actors. The survey was conducted at a time when the long ruling *Partido Revolucionario Institucional* (PRI) had just lost the elections for the first time in 71 years and the new administration had replaced many of the political actors formerly involved in the biotechnology debate.

It is clear that there are difficulties in addressing the dynamic aspects of a public debate through surveys. However, interpreting the results of these surveys in the context of a particular political system and its cultural and historical background, may help to explain future political events and changes in perception and political influence of old and new political actors.

The aim of the overall research project was to understand the knowledge, perceptions and interests that shape debates on agricultural biotechnology in developing countries. The two surveys conducted in the Philippines and Mexico indicate that expectations and concerns about agricultural biotechnology may differ from those prevailing in developed countries. Moreover, a comparison of stakeholder perception in Mexico and the Philippines shows that the political, cultural and historical backgrounds significantly influence the issues that are discussed at the national level. The conclusions indicate that a global system of governance of biotechnology has to take into account perceptions specific to individual developing countries, not only to manage risk, but also to ensure the fair and equitable distribution of benefits.

Regulations, Negotiations and Campaigns: Introducing Biotechnology into India

Criticisms and Doubts: Biotechnology has been received with a considerable degree of scepticism in Indian agricultural circles. Some critics questioned whether it could make any significant contribution to increasing food security, while others warned that it might actually threaten food supplies by stimulating the production of cash crops. There was also the possibility that it might adversely affect agricultural exports by encouraging the developments of synthetic substitutes for commodities such as sugar, an important Indian export.

Biotechnology was also associated with the private sector becoming increasingly involved in the provision of seeds and other planting material. Government policy in India is committed to ensuring food security. Public sector research focuses on both food and commercial crops and takes a balanced view of agriculture as a whole. The private sector, however, is primarily concerned with commercial crops and has largely neglected the orphan crops that form the mainstay of the diets of small and marginal farmers and agricultural labourers. The growing power of the private sector is seen by some to be undermining food security objectives.

Genetically Modified Crops

It was the issues associated with terminator technology that provided a focus for concern about India's agrobiotechnology policy. Those involved in the ensuing debate came from all sectors of Indian society and included policy makers, providers of technology from publicly funded and private sector organisations, civil society groups and farmers organisations.

In 1998, *Delta and Pine Land Company* (USA) and the *United*

States Department of Agriculture (USDA) obtained a patent for a new genetic technology designed to produce sterile seeds. These seeds, soon dubbed 'terminator seeds', were intended to prevent farmers from reusing seeds obtained at harvest and were expected to increase the value of the proprietary seed owned by USA seed companies and open up new overseas markets.

India's connection with terminator technology came through the life science company *Monsanto* (USA) who after acquiring Delta and Pine Land declared its intention to market *genetically modified* (GM) seeds despite opposition from the Green lobby. The company's decision to acquire terminator technology provided the catalyst for the launch of a determined campaign against GM seeds by a coalition of civil society organisations and farmers' groups in India.

Many of these groups were already lobbying against the decision to introduce *Intellectual Property Rights* (IPRs) into agriculture. In the multilateral negotiations that accompanied TRIPs and the Cartagena Protocol on Biosafety, the Indian Government had not only played an active role, it had also taken positions at critical moments in each set of negotiations. These positions were strongly influenced by the campaigns launched by those civil society organisations that sought to highlight the interests of Indian farmers and the need to protect the environment.

The key issue in meeting the commitments that the Indian Government had made under TRIPs to the *World Trade Organisation* (WTO) was the type of IPRs that should be introduced. One of the options open to WTO members was the adoption of the framework provided by the *International Union for the Protection of New Varieties of Plants* (UPOV). The UPOV *Convention,* formalised in 1961, provided a system for granting Plant Breeders' Rights. In the past, the UPOV Convention had allowed exclusive rights to be granted to plant breeders for the varieties they had developed, and at the same time it also allowed traditional farmers to reuse seeds obtained from their annual harvest and to exchange these seeds with neighbours. The most recent amendment to UPOV, carried out in 1991 (UPOV,91), however, took away the freedom granted to farmers to exchange seeds with their farm neighbours or re-use seed for commercial purposes. UPOV,91, therefore, took away what farmers saw as one of their basic rights.

Small farmers in India had been vocal in their protest against UPOV,91 and the Indian government did not adopt the UPOV option. The discussion did, however, contribute to heightening farmer suspicion. It was against this background that the idea of terminator seed entered India. It was, therefore, no surprise that the farming community forthrightly rejected it. The government's firm position against the introduction of terminator seed was clearly influenced by civil society campaigns that were determined to prevent a strengthening of intellectual property laws that would make it easier for seed giants to dominate the local seed market.

Genetically Modified Seeds

The controversy over the threat from terminator seeds had considerable fallout, jeopardising field trials of a GM variety of cotton that Monsanto had been conducting since mid-1998. This variety of cotton had been developed by implanting a gene extracted from the naturally occurring Bt soil bacterium into the high yielding varieties of cotton already being cultivated in India. Monsanto claimed that this new variety of cotton would improve local cotton varieties and provide them with the potential to kill insect pests.

This would result in substantial yield increases and, equally importantly, would offer what was presented as a more environmentally safe way of cultivating a crop that currently accounted for 40 per cent of the total volume of pesticide used in Indian agriculture each year.

In 1998, the Indian government had given the go-ahead for limited field trials of Bt cotton. Its view was that Monsanto's cotton variety could provide the necessary resistance to the bollworm pest and help prevent annual losses estimated to be worth up to US$ 190 million dollars. The *Indian Council of Agricultural Research* (ICAR) justified support for the adoption of Bt cotton on the grounds that the conventional crop required heavy applications of chemical pesticide in order to flourish. ICAR argued that conversion to Bt cotton would not only reduce environmental pollution, it would also reduce the burden of debt farmers were often forced to incur in order to purchase the pesticide needed to treat their cotton crop.

However, one of the main points raised by those campaigning against the decision to introduce Bt cotton into India was that the process for approving field trials had not been carried out with the requisite level of transparency about the environmental risks of using GM seeds.

Regulations

The regulation of *genetically modified organisms* (GMOs) in India has been subjected to the rules framed by the *Ministry of Environment and Forests* (MOEF) in 1989. These rules, which were part of the *Environmental (Protection) Act* of 1986, defined implementing structures for conducting research and for the commercial applications of GMOs. This regulatory structure consists of six committees: three are under the *Department of Biotechnology* (DBT), one under the MOEF and two operate at subfederal levels closer to the actual site of GM crop field trials. However, although the Government had established this comprehensive structure for regulating GMOs as early as 1989, long before the global community reflected on the need to pursue biosafety regulations, critics in India have raised questions about the way the system operates.

One of the more pertinent questions relates to the fact that this structure of six committees is largely redundant since only two of the committees seem to have any role to play. These are the *Review Committee on Genetic Manipulation* (RCGM) of the DBT and the *Genetic Engineering Approval Committee* (GEAC) of the MOEF. The functions of these sub-federal committees remain largely undefined, and the way in which rules have been framed for the committees concerned with the regulatory administration of GMOs have given cause for concern.

The issue of Bt field trials brought these concerns to the fore. The regulatory procedure followed by the RCGM in the case of Bt cotton was questioned on several grounds by civil society organisations. First, field trials were approved by the RCGM after the farmers chosen for this purpose had sown the Monsanto seeds in their fields. Second, there was criticism of the role played by the sub-federal committees in granting approvals for field trials at 40 locations in nine states. Finally, the functioning of RCGM was called into question because changes in the rules of the committee were made when Bt trials were still in progress.

This lack of transparency in granting clearance to the field trials of Bt cotton was the single most important factor behind the farmers' protests in the Southern states. Proponents of genetic engineering argued that civil society campaigns had spread misinformation and re-emphasised that the objective of promoting biotechnology was essentially to increase the viability of Indian farmers. Misinformation or not, the campaigns managed to put sufficient pressure on the government to prevent an early introduction of Bt cotton on the market.

Government Decision-making

The way in which Bt cotton was introduced into India illustrates the complex nature of the processes that are involved in governmental decision-making. In the past, vital decisions were often made and implemented before their implications were analysed. Today, we are witnessing a more active involvement of stakeholders before decisions are made on any particular issue. Many would insist that the lack of transparency still pervades but times are in fact changing.

In 2000, the three Southern Indian states of Karnataka, Andhra Pradesh (both scenes of farmer protest in 1998 and 1999) and Tamil Nadu gave explicit support for biotechnology and began to announce their biotechnology policies. This has provided the necessary mandate to the federal government to fulfil its unfinished task of completing the Bt cotton field trials. State level support for biotechnology has a deeper significance in the Indian context where the Constitution of India (which delegates powers between the central government and the states) defines agriculture as one of the issues that should be managed exclusively at state level.

It is becoming clear that the number of seed companies and State governments in favour of introducing GM seeds is increasing. The seed companies are anxious to introduce GM seeds as part of their corporate strategy, while state governments are more inclined to see these products in the context of their stated policy of modernising agriculture. In July 2000, the Genetic Engineering Approval Committee of MOEF approved large-scale field trials of Bt cotton by the Monsanto affiliate Maharashtra Hybrid Seeds Co. Ltd. (Mahyco) for seed production and demonstration and to generate environmental safety data for crops under various agricultural and climatic conditions. Mahyco was to undertake open field trials on 85 hectares and seed production on 150 hectares. Most significantly the ICAR, which includes agricultural universities as well as public-funded research institutions, would be fully involved in monitoring seed production. The decision to involve the ICAR system was taken even as the Central Institute for Cotton Research, one of the public-funded agricultural research institutes, was unveiling plans for putting its own variant of Bt cotton on the market by the year 2002.

The domestic biotechnology industry was beginning to assert itself in India, after spending years in the shadow of corporates like Monsanto. A start was made in April 2001 with agriculture high on the agenda and crops such as cotton, rape and mustard under research.

However, potential users of GM seeds within the farming community as a whole still do not appear to be convinced of their potential. In December 2000, the Minister of Agriculture informed the Indian Parliament that a final decision on the use of GM seeds had yet to be taken. The campaigns run by civil society organisations have certainly contributed to this decision but much more coordinated effort will be required if these concerns are to be translated into policy.

Ethics: the Key to Public Acceptance of Biotechnology

Although some biotechnologists think that bioethics is a product of the 1990s, it actually has a much longer tradition. As with medical ethics, professional bioethics traces it roots to early ethical codes of conduct such as the Greek Hippocratic oath. This was the first explicit acknowledgement in the 'life sciences' of professional responsibility requiring that physicians should not harm their patients.

It was, however, only in the 1960s that bioethics started to move beyond the classical medical ethics issues of doctor patient relationship. At that time bioethics developed into a coherent professional discourse and discipline, encompassing a variety of ethical issues raised by progress in medical science and technology. Moreover, it slowly began to encompass other ethical issues, such as our relationship towards animals and the environment. It is in this broad sense that the term bioethics is currently used, although medical ethicists still like to reserve it for (human) medical ethics alone.

A number of important developments in society and science can be identified as having led to proliferation within bioethics and the continued expansion of its sphere of influence. Bioethical questions could, for example, only be raised in the context of a secularising world. In Western society, the moral crisis that was the consequence of the loss of the Biblical viewpoint as the supreme moral authority, gave rise to an ethical and metaphysical fragmentation that forced society to face and discuss bioethical questions. Much later, in the second half of the last century, an increasing awareness of the vulnerability of fauna and flora further supported the already growing ethical status of non-human animals and the environment. It helped clear the way for the recognition that non-human animals are also worthy of moral consideration for and by themselves, and even that species and ecosystems are deemed morally relevant for their own sake.

As a consequence, a broad variety of human practises which traditionally were regarded as 'normal' suddenly became morally questionable and the subject of societal debate. This article however, will focus on the theme of this edition of the *Biotechnology and Development Monitor* and specifically examine how the dynamics in the relationship between science and society and public perceptions of science have contributed to the present ubiquity of bioethics in public debate.

Dimensions of Risk

Scholars in the social sciences have written countless critiques on the claims of science to objectivity and certainty. These critiques have not only dealt with these issues in general. Social science scholars have written extensively on the psychological and social construction of risks, and also on specific issues such as sound science stance on biotechnology risk management.

Probably the most influential sociological critique on how risks are generated and handled in industrialised, post-modern societies, has been written by the German sociologist *Ulrich Beck*. In his book, *Risk Society* he convincingly demonstrates the political nature of risks caused by technologies in post-modern industrialised society. He argues that in our times, the production of wealth ('goods') is systematically accompanied by the production of risks ('bads'). Modern risks, like ecological disasters and nuclear fallout, for example, typically inflict irreversible harm and are only indirectly visible, based on (scientific) causal interpretations. Since the consequences of these risks can have global effects, they render national jurisdictions inadequate and harm even the rich and powerful. They, therefore, also raise the issue of democratic accountability and control since, as Beck puts it, *"safety issues are illegitimately decided by the corporations of engineers of our high-risk civilisation, under cover of the empowerment formula—state of science and technology."*

Once these 'late modernity' risks and their societal consequences were recognised—which is now the case in Western societies—what was once considered to be apolitical becomes political. Hence, the risks engendered by biotechnology—traditionally approached as if they were an objective, scientific issue—have turned into intense political debate. In Europe this has lead to biosafety policies in which ethical political concepts like the *precautionary principle* demand that both scientists and regulators apply a new approach - explicitly moral in its influence - to biotechnology risks.

That Beck's analysis is not merely one academic's view of society is repeatedly demonstrated in public opinion polls such as the European Commission's Eurobarometer, which has been regularly surveying public attitudes to biotechnology since the late 1980s.

Surveys such as these show that an acceptance of exposure to risks depends on whether an application is considered useful and morally acceptable. Moreover, moral concern acts as a veto regardless of perceptions of risk and use. The conclusion which has been drawn from this is that if risk itself is less significant than moral acceptability in shaping public perceptions, then public concerns are unlikely to be alleviated by technically based assurances and other policy initiatives that deal solely with scientific risks. It is this hard lesson that has finally caused industry, regulators and politicians, to face the fact that the ethical dimension of risk, and ethical issues in general, lie at the root of the societal controversy about biotechnology.

Professional Ethics

The persistent societal unease with biotechnology and increasing awareness of the influence ethics exerts on public perceptions has helped the biotechnology community to realise the importance of communicating a concern for responsible 'bioethical' conduct to society. This became all the more critical in the mid 1990s, when for the first time, industry had to convince consumers to support biotechnology by buying its genetically modified harvest.

It was at this time that both professional codes for individual biotechnologists, such as the European Federation of Biotechnology's code of conduct, as well as codes for bioindustry organisations, such as BIO's Statement of Principles, were developed. In professional circles it is not uncommon to 'organise' an ethical position in the form of an, often voluntary, explicit professional code of conduct. These codes traditionally have both internal and external functions. Internally, they provide the means to build a professional ethos, improving the quality of the services delivered to clients. Externally they also serve as a means to build trust with society. They are the 'certificate' that shows that the profession as a whole will act responsibly towards both clients and society.

Opinion surveys in Europe have repeatedly shown that biotechnology suffers from a lack of public trust and acceptance of its products. This is one of the principal reasons why the biotechnology community has recognised the importance of having a professional ethical code.

Often a professional code is also a way for the sector to stay one step ahead of compulsory, and usually more restrictive, codes; namely governmental laws and regulations. This second external function is one of the reasons why such codes have become popular with both industry and regulators. They provide mechanisms for a self-regulation that is less consumptive of time and resources and which is therefore, usually preferred by both sides. However, general these codes may sometimes be, their consequences are far from trivial. Just the creation of a general ethos can have far-reaching consequences. By formulating such codes the biotechnology community as a whole has acknowledged for the first time that there are more perils to manage in biotechnology than mere technical risks. In doing so, the biotechnology community itself has helped to put bioethics on the public agenda.

It should also be recognised that typically these codes do not restrict themselves to merely stating an ethical respect for patients, animals and environment. They often go further by explicitly acknowledging the fact that biotechnology does indeed raise social issues that requires industry to respect the regulatory ideals of transparency, accountability and public participation.

Thus, these codes themselves focus the spotlight on all sorts of socio-economic and socio-cultural consequences only indirectly tied to science and (bio)technology. It is the types of 'globalisation' issue that have evoked the fiercest protests in the streets and have been hardest to put on the agenda of international bodies such as the World Trade Organisation (WTO) and the Organisation for Economic Cooperation and Development (OECD).

By putting biotechnology in the context of globalisation, the societal debate on biotechnology inevitably had to shift further towards discussing the ethical and social impacts of biotechnology.

Future of Bioethics

As the official definition of bioethics formulated by the International Association of Bioethics makes clear, the wider social context within which ethical issues are being increasingly defined, such as globalisation, now provides the context for bioethical enquiry. The Association defines bioethics as the study of the ethical, social, legal, philosophical and other related issues arising in health care and in the biological sciences. With such a broad and general definition, bioethics and its practitioners are likely to show up anywhere stressing the importance and prominence of their discipline. This broadness, however, is also a cause for concern. Bioethics is vulnerable to the criticism that it is *"all talk but no action."* The interest of industrial stakeholders in bioethics is largely based on the hope that it can guide society into a bioindustry safe haven where global consumer acceptance of biotechnology prevails. Whether bioethics has this capacity remains to be seen. Both within and between societies radically different appreciations of particular bioethical values can be identified.

Taking an autonomy based rather than a community based approach, or a rational reductionist, rather than a spiritual holistic approach to ethics, are just two of the many differences to be found across the North-South and East-West divide. Considering its pluralistic nature, bioethics should perhaps focus more on guiding society to live more peacefully with moral diversity and it is very possible that this will become the dynamic that will drive the discipline forward in our immediate future.

Environmental and Industrial Biotechnology in Developing Countries

Modern biotechnology, which refers to the application of biological organisms, systems and processes to the provision of goods and services (OECD, 1998), has grown from being solely the domain of research scientists in academic laboratories to being an enabling technology capable of being incorporated into many different sectors of an economy.

Much of this development has thus far been concentrated in the medical field and involves drug-directed research by pharmaceutical and biotechnology companies. The second largest application, although only one-tenth the size of the medical sector in terms of funds spent and income generated, is in agriculture (Kate and Laird, 1999). Interestingly this latter sector has captured the public spotlight to a much greater degree than its larger counterpart mostly as a result of the actions taken by domestic and international interest groups. The third broad sector of biotechnology is in its application in industry and to environmental problems. So far this sector has received a lack of attention not only from the media, but also the financial and policy perspectives. This paper is intended to give an overview of the opportunities and barriers to the industrial and environmental application of biotechnology in developing countries.

The target areas for industrial biotechnologies can be seen as falling into three main categories: industrial supplies (biochemicals, enzymes and reagents for industrial and food processing); environmental (pollution diagnostics, products for pollution prevention and bioremediation); and energy (fuels from renewable resources). Industrial processing refers to chemicals, pulp and paper, textiles and leather, while environmental applications relate mostly to the effects of the metal and mineral industries. In the energy sector, an important area is the replacement of fossil fuels with renewable raw materials, (OECD, 1998; Kate and Laird, 1999).

In assessing the potential of industrial and environmental biotechnology in developing countries, some points should be considered. First, even in industrialised countries, the expected economic benefits of this sector of biotechnology are orders of magnitude smaller than those found in pharmaceuticals and agriculture (OECD, 1998; Kate and Laird, 1999).

This essentially restricts activity to niche markets and justifying production changes with environmental considerations. Second, the capital outlays required to use bio-based industrial processes are presently higher than those of traditional mechanical or chemical processes.

Thus, it can be difficult to justify starting new industries based on biotechnology, which leaves such innovations to existing facilities. This leads to the observation that only intermediate developing countries, those with existing industries and some scientific capacity, can realistically consider applying biotechnology to industrial processes. Other developing countries could still become involved in this sector in a limited fashion, through the use of their biological resources.

From the economic perspective, early studies in industrialised countries indicate that replacing existing chemical-based production methods can lower operating and energy costs (Griffiths, 2001). This report also found a consistent reduction in hazardous waste products.

Thus, the industrial application of biotechnology, if managed properly, could have a net positive effect in developing countries. The qualifier "net" is an important one in this context. Some potential negative effects arise from the Schumpterian concept of innovation, which recognises that a new innovation can render its predecessor obsolete while creating new opportunities. The potential negative consequences of obsolescence, such as worker displacement, dependence on new materials and changes in process management, should be considered to avoid running the risk of raising expectations too high, as has been the case with agricultural biotechnology.

The incorporation of biotechnology into industrial settings in intermediate developing countries also faces financial barriers. Largely, this comes down to not having enough money for capital equipment upgrades. Since bank financing is not an option in many developing countries (with banks demanding collateral of equal value to the loan), private investment, either domestic or foreign, must be considered. Here there are several constraints including domestic institutional gaps, such as lacking private venture capital companies or underdeveloped stock markets to liquidate investments. There is also missing public capital investment resulting from economic stagnation and falling commodity prices (Avramovic, 1996).

On the positive side, it may be less expensive to overhaul an obsolete factory in an intermediate developing country than to do the same in an industrialised one. Thus international investors or companies may be more willing to consider bio-based process for their industrial partners in developing countries. Another advantage is that many industrial technologies are scale-neutral so that even small businesses can take advantage of them. Interestingly, the special challenges in obtaining international financing of industrial biotechnology in developing countries may be smaller in some respects than for health or agricultural biotechnology. This derives mainly from the consideration of mining, pulp and paper, textiles and chemical manufacturing as traditional investment targets since investors can calculate net present value and other accounting necessities based on known valuations and experience.

Structural resistance can also hinder the successful introduction of biotechnology into current industrial enterprises. One source of resistance can come from chemical and mechanical engineers with no biological training or even knowledge of the potential benefits of biology to their trades.

To them there is no obvious advantage to change processes. To effectively manage this inherent resistance, there needs to be a local pool of scientists that can explain the role of biotechnology in the manufacturing process participates during the technology transfer process. Another potential source of structural resistance comes from the nature and context of the targeted industry.

The end-users may not require that an entire process is overhauled, only one or a few steps. Any involvement by a third party in trying to assist in biotechnology transfer should consider the absorption capacity of an industry with regards to fundamental research breakthroughs. It has been noted that investments in upgraded equipment can be wasted, unless accompanied by investments in complementary methods of organisation and management (Milgrom and Roberts, 1990). Furthermore, in order to be used effectively, processes involving biotechnology do require certain production standards. It is important to bring industrial processes involving biotechnology up to international standards from the beginning. Not only will this provide the option of exporting products, but it will also prevent unexpected future capital outlays should domestic regulations be rewritten to accommodate international obligations.

Taking the aforementioned conditions into consideration,

it seems that the most likely form of industrial biotechnology that could be introduced into intermediate developing countries in the near future is in cleaner production processes. This type of technology introduction would only need to affect specific parts of a given production process, could show economic benefits relatively quickly and also help companies conform to environmental obligations.

One factor that should be remembered is that often "dirty" industrial processes arise from the tendency of foreign investors to engage in ventures with high yields in the short term so as to avoid economic and political volatility (Newell, 2001). Thus it is often foreign-owned firms that need to be convinced of the value of cleaner production. A useful contention could be that biotechnology is not "clean" *per se* but that environmental and economic sustainability flow from producing wastes that are degradable or recyclable (OECD, 1998).

With respect to the direct environmental application of biotechnology, there are options as well. An example of replacing a chemical process with a bio-based one comes from Kenya and the product BIOFIX, which is an alternative to chemical fertilizers. Measurements show that 100g of BIOFIX replace 90kg chemical nitrogen (Zechendorf, 1999).

This reduces not only toxic environmental exposure but human as well since fertilizers are spread manually in most developing countries. In the energy sector, Brazil is developing a strategy for using soy oil and sugarcane as a source of ethanol to be refined into biodiesel (Ford, 2000).

Realistically though, any effort to promote bio-based substitutes for petroleum products will have to be undertaken specifically to support either a particular primary production sector or environmental directives since oil is currently available at the lowest price in history. Other uses of biotechnology in the environmental sector relate to the production and distribution of substances that absorb toxic products from either soil or water. Most of these processes are not yet economically viable in industrialised nations and so it would be difficult to transfer them to developing countries at this time.

The only viable involvement for many developing countries at the present time in industrial or environmental biotechnology is based on providing the products of natural resources, such as enzymes, as process components. Inherently, due to lacking infrastructure and financial resources, this involvement will require foreign investment and/or partnerships that would have to navigate through the international debates surrounding benefit sharing and access rights.

Intellectual property rights are currently and will continue to be for the foreseeable future, a relatively unimportant factor in the transfer of industrial and environmental biotechnology. For example, in the pharmaceutical sector, an empirical study has shown that patents and patent law are not necessarily the major barrier to AIDS treatment across Africa as has been assumed by many in the development field (Attaran and Gillespie-White, 2001). In many developing countries, economic barriers alone will suffice to allow near exclusive financial benefits since the capital outlay for implementing biotechnological processes cannot be easily duplicated locally. If monetary conditions do eventually allow for competition, then an industry has been established and local players will have both the means and interest to demand appropriate legislative action from their government.

Policy Directions

Some factors appear when considering the most effective means for international institutions to help introduce biotechnology into the industrial sectors of the developing world. First, success should be demonstrated by targeting projects to appropriate regions and existing industries. One useful concept is that of intermediate developing economies, which have a scientific base but weak industrial structure. Such economies often exhibit a dependence on imported technology and a limited degree of industrial integration.

However, the basic ingredients are present, so new initiatives can have a noticeable impact upon the status quo by demonstrating the viability of biotechnology. As well, since even a small scientific community will be familiar with the techniques being introduced, a local resource exists for helping other entrepreneurs adapt to the new processes.

This could start a "virtuous cycle" (Fontes, 2001). Another factor to consider involves relating capacity building exercises to international environmental agreements. The use of biotechnology in implementing cleaner production methods, especially in industries with heavy chemical use such as leather and mining, could be more actively assessed and promoted.

In addition, international agencies can promote private sector involvement by helping mitigate risk through the establishment of mechanisms to identify promising technologies and perhaps even provide parallel investments at the initial stages of process conversion or technology development (Bazley, 1999). Microfinancing for small company owners is another option (Walsh *et al.* 1995). The above strategies could also be applied to supporting private sector investment in identifying and characterising natural resources, such as enzymes, that could be used in industrial processes.

Strong local government involvement is needed, if not in large-scale structural and financial support, then at least by publicly displaying a positive attitude to both biotechnology and entrepreneurship (Bazley, 1999). Governments could do well to remember that biotechnology is an enabling technology because of its widespread impact. Many process/enabling technologies may increase the productivity and sustainability of industrial processes currently in use and as such, major technological changes can initiate structural changes and economic growth across an industry.

From the standpoint of the companies that could consider biotechnological processes, one important issue is the economic advantage gained from entering or creating a particular niche market. Another important issue would be the possibility of reducing pollutant emissions at the source. This may not be of direct economic benefit since the majority of developing countries have either weak or non-existent enforcement of environmental regulations regarding waste treatment, however it may be used to enhance public image and generally improve the environmental conditions of the workforce.

Since economic benefits from industrial and economic biotechnology will be relatively small in any case, it may be useful to think about the use of a highly publicised subsidy structure as an incentive to manufacturers to adopt new technologies. This structure may take place in the form of guaranteeing large-scale orders (Mehra, 2001). Such schemes are used in the US and Brazil to promote alternative technologies such as ethanol fuel and bioreactors to benefit their agricultural sectors.

In the near future, industrial and environmental biotechnology will likely only benefit those intermediate developing countries that have already established some industrial enterprises, for example in leather or metal production, and have at least a rudimentary set of environmental regulations.

Reasons for this include the reality that it is generally easier to modify existing industries than to create new ones and that even minimal pollution control measures can be used to convince industrialists to take advantage of external assistance, either national or foreign, to adopt bio-based cleaner production process.

New Biotechnological Applications of Coconuts

Plants are the best choice for the production of recombinant proteins (Ma *et al.* 2003). Three principal methods are used to produce these proteins, all of which require a DNA sequence to be incorporated into the plant genome: (1) infection with *Agrobacterium tumefaciens* (Gelvin, 2003), (2) electroporation (Trezise, 2002) and (3) biolistics (Taylor and Fauquet, 2002).

To date, no report has appeared on the capacity of coconut water (a syncytial endosperm tissue) to produce recombinant proteins. This paper shows with simple fluorescent EGFP microscopy, that coconut water expresses foreign DNA sequences. Since hundreds of milliliters can be extracted from each coconut, it follows that coconuts can be exploited for large-scale, industrial production of recombinant proteins using bioreactor technology. Furthermore, as coconut water nuclei are naked, devoid of cell membranes, the micropropagation of genetically engineered plant clones is greatly facilitated.

Coconut Water

From each coconut tree, a single coconut was used if it had a liquid syncytium containing more than 10^5 nuclei per milliliter. This corresponded to 4-6 months of development. To avoid bacterial contamination, the coconuts were studied within 5-15 m after picking.

The coconuts were gently removed with sharp scissors. As higher temperatures accelerated expression, the coconuts were removed from the tree at 12:00-14:00 hrs (*i.e.* the hottest time of the day). As a consequence, experiments were not carried out during cold and rainy days. Coconut water was extracted under sterile conditions.

A single transversal cut was made at the top of the coconut with a sharp, 10 inch-long commando knife. The water was then extracted with a 10 cm-long, 20 gage Teflon needle (Sigma) attached to a 60 ml syringe (B-D Biosciences).

Metal was avoided because the colourless, transparent coconut water turned brownish within 5 minutes upon contacting metals. This response of the coconut water to metal contact seems to be caused by the low pH of the coconut water (pH 4.0-5.0 - Bustamante, 2002a). That the metal blade of the knife did not produce any reaction was confirmed visually by the stability of the water transparency and lack of colouring. Sterile conditions were maintained with UV light, with HEPA air filtration, with boiling water sterilisation, and with anti-bacterial sterilising liquids. Caution was taken to prevent coconut particles from falling into the water (during and after cutting the top of the coconut) as this brought about browning of the water which was interpreted as a negative reaction of the living syncytial endosperm.

Light Microscopy

The coconut water was studied with a custom-built inverted microscope consisting only of a 100X objective and a 15X eyepiece (Bustamante, 1991; Bustamante, 2002a). The light from a mercury-xenon lamp (Oriel) powered with a stabilised voltage source was used in conjunction with a dichroic filter set for EGFP (Omega Filters). An inexpensive CCD webcam camera (Quickcam Pro P3000 and P4000, Logitech) was used to detect and measure the fluorescent light level. The camera was attached to the eyepiece. The experiments were carried out in a 100 μl chamber with a bottom made out of a microscope coverslip. Additional information on the experimental system has been published (Bustamante, 2001; Bustamante, 2002a; Bustamante, 2002b). Experiments were carried out at room temperature (24-26°C).

Recombinant DNA Vector Specifications and Method of Application

The plasmid pEGFP-C1 (Living Colours Family of Vectors, B-D Biosciences-Clontech) having a mass of ~3.1 MDa, was selected for the testing of the capability of coconut water to synthesise recombinant proteins. The plasmid stock solution was prepared in the plasmid original commercial storage vial by adding fresh coconut water filtered through a 0.22 μm syringe filter (199-2020, Nalgene). Unused portions of the stock solution were transfered with DNAse- and RNAse-free pipette tips (*Biopur*, Eppendorf) into DNAse- and RNAse-free microcentrifuge tubes (Biopur, Eppendorf) and stored at -80°C. A small volume (1-5 μl) of the plasmid stock solution was directly applied to the experimental chamber containing freshly extracted coconut water (*i.e.* liquid syncytial endosperm).

The pipette used for application of the plasmid was also used to gently stir and thus dissolve the added plasmid into the final 100 µl of working coconut water. The final concentration of the plasmid varied between 1 and 100 pM. Only results for 100 pM are shown. Imaging with the inverted microscope was carried out through the coverslip that formed the bottom of the experimental chamber. The fluorescence level was calibrated following standard procedures (*i.e.* Liu *et al.* 2001; Ehrhardt, 2003).

Twenty four coconuts of 93±12 mm diameter, 111±18 mm length and 149±37 ml (mean±SD) were used. The cell-free/ naked nuclei had internal and external movements.

Syncytial endosperm contains cell-free, naked nuclei that continuously divide because one of the functions of the nuclei is to form new layers of endosperm cells (Cutter and Freeman, 1952). Fourteen divisions were observed in all the coconuts studied. These divisions took about 1 hr to complete. The slow division could be characteristic of the 12 months that take for coconuts to reach maturity. Another possibility is that the slow division may result from changes in the chemical and/or physical conditions. The observed density and the mechanical activities directly correlated with the level of expression of EGFP.

This observation may be a good assay for *in loco* coconut selection. Unmolested, 'virgin' trees were the best trees for EGFP expression (r.u. 98±6 per cent, n=8). Trees disturbed weekly (one coconut picking per week) had only slightly lower expression levels (r.u.=89±9 per cent, n=16). Trees that were allowed daily disturbances reduced their nuclear density and nuclear movements within one month, suggesting a 'programmed' hibernating defence mechanism. The entry of pEGFP-C1 into the naked nucleus, exclusively via the nuclear pores, is assisted by the nuclear localisation signals present in the transcription factors attached to the plasmid (Bustamante, 2002a). For bioreactor technology this may not be relevant. However, this greatly facilitates plant genetic engineering and clonal micropropagation. The large volume of water in each coconut translates into significant savings because no culture broth is needed (as is the case for microbial fermentation systems).

In regard to micropropagation, one could question the fact that the endosperm is triploid tissue. However, this should raise no concern because it has been demonstrated that hexa- and dodecaploid plants produced from endosperm (Cantoni *et al.* 1993; Tao *et al.* 1997; Chaturvedi *et al.* 2003).

The demonstration that an inexpensive webcam CCD camera (about US$100) can be used to measure protein expression in coconut seems significant to the dissemination and implementation of the concepts presented here. The success of the low-tech approach for the detection of EGFP synthesis by coconut water demonstrates that the level of expression is indeed high. The success of this simple approach results from at least three facts.

First and foremost, as shown here, coconut water is an excellent system for recombinant protein expression. Second, the simplified design of the custom built microscope minimises the optical path and, therefore, optimises the light-gathering capacity of the detection system. Third, the stability of EGFP results in the continuous integration of the signal over time. The use of coconut for the production of expensive substances (either through bioreactor or genetic engineering for clonal micropropagation) should be of interest to poor nations. Coconuts should prove an excellent refrigeration- and needle-free system for the delivery of nutriceuticals and therapeuticals to the peoples of all nations and, specially, to the peoples of poor developing countries.

Isolation and study of a ubiquitously expressed tomato pectin methylesterase regulatory region. Pectin methylesterase (PME) is an enzyme that have been found in every plant tissue analysed (Lineweaver and Jansen, 1951; Rexova-Benkova and Markovic, 1976), several fungi (Christgau, *et al.* 1996;

Mendgen, *et al.* 1996), bacteria (Plastow, 1988; Barras *et al.* 1994) and even insects (Ma *et al.* 1990; Shen *et al.* 1999). In higher plants, it is known to be a cell wall associated protein and several of the PME cDNA available in the literature, are known to have towards the N-terminal sequence, a characteristic signal peptide which is thought to help in targeting the protein to the plant cell wall (Gaffe *et al.* 1997 and references therein). PME catalyses the deesterification of galactosyluronate methylesters of pectins, releasing protons and methanol into the media (Frenkel *et al.* 1998). Despite the biochemical mode of action of PME is well known, it have been difficult to demonstrate any role for PME in the physiology of plants. However, several hypothesis had been proposed: pollen germination and/or tube growth (Mu *et al.* 1994), abscission (Sexton and Roberts, 1982), regulation of cell enlargement through changes in the plant cell wall Donnan potential (Ricard and Noat, 1986), fruit softening during post-harvest fruit ripening (Zeng *et al.* 1996) and plant defence (Chamberland *et al.* 1991; Wietholter *et al.* 2003).

Furthermore, strong experimental evidences had been provided to suggest that a PME role in the release of cells from the root cap (Stephenson and Hawes, 1994; Wen *et al.* 1999), plant pathogenesis (Collmer and Keen, 1986; Mendgen, *et al.* 1996; Nun *et al.* 1996; Valette-Collet *et al.* 2003), plant systemic infection by tobacco mosaic virus (Chen and Citovsky, 2003) and maintenance of the tomato fruit tissue integrity during post-harvest shelf life (Tieman and Handa, 1994). However, the actual physiological role of PME is still matter of controversy. Four papers to our knowledge had been published in which the cloning of regulatory sequences of PME from higher plants was described. However, in all of them DNA comparison by computer was the only tool used to prove that the gene located downstream was indeed encoding a pectin methylesterase. Albani *et al.* (1991) reported the finding of a genomic clone from *Brassica napus* which contains the PME gene and its 5′ upstream regulatory region.

Studies were conducted using a piece of the gene located dowstream as a probe. This gene was found to be expressed mainly during pollen development.

Two putative PME promoter regions were cloned from *Brassica campestris* (Kim *et al.* 1997). Study of their sequence found them to have high homology with the previously reported promoter PME from *Brasicca napus* (Albani *et al.* 1991). Further, a sequence motif similar to the one known to exist in two tomato pollen-specific promoter was located.

Tobacco transgenic plants with constructs containing two different promoter sizes from one of those two promoter available were made. Expression of the GUS gene was only detected during developing and mature pollen grains germinated in vitro. Recently, the cloning of two 5′ upstream region of PME from *Citrus sinensis* was published. Northern blot analysis showed that both DNA regulatory region are active in most of vegetative tissues (Nairn *et al.* 1998).

In our laboratory, we have cloned a 13.7 kb. genomic DNA from tomato containing a 2.59 kb. of DNA 5′ flanking region, along with all the PME genomic clone.

Identification of the protein encoded by the gene downstream was made by creating tobacco transgenic plants over expressing the PME cDNA (Gaffe *et al.* 1997) and comparing the sequences of the genomic and cDNA regions. In this work, we describe the study of a 5′ flanking region of a PME gene called pmeu1 (which stands for PME ubiquitous one) using computer tools and tobacco transgenic plants.

Pectin methylesterase (PME) is an enzyme that have been found in every plant tissue analysed (Lineweaver and Jansen, 1951; Rexova-Benkova and Markovic, 1976), several fungi (Christgau, *et al.* 1996; Mendgen, *et al.* 1996), bacteria (Plastow, 1988; Barras *et al.* 1994) and even insects (Ma *et al.* 1990; Shen *et al.* 1999). In higher plants, it is known to be a cell wall associated protein and several of the PME cDNA available in the literature, are known to have towards the N-terminal sequence, a characteristic signal peptide which is thought to help in targeting the protein to the plant cell wall (Gaffe *et al.* 1997 and references therein). PME catalyses the deesterification of galactosyluronate methylesters of pectins, releasing protons and methanol into the media (Frenkel *et al.* 1998).

Despite the biochemical mode of action of PME is well known, it have been difficult to demonstrate any role for PME in the physiology of plants. However, several hypothesis had been proposed: pollen germination and/or tube growth (Mu *et al.* 1994), abscission (Sexton and Roberts, 1982), regulation of cell enlargement through changes in the plant cell wall Donnan potential (Ricard and Noat, 1986), fruit softening during post-harvest fruit ripening (Zeng *et al.* 1996) and plant defence (Chamberland *et al.* 1991; Wietholter *et al.* 2003). Furthermore, strong experimental evidences had been provided to suggest that a PME role in the release of cells from the root cap (Stephenson and Hawes, 1994; Wen *et al.* 1999), plant pathogenesis (Collmer and Keen, 1986; Mendgen, *et al.* 1996; Nun *et al.* 1996; Valette-Collet *et al.* 2003), plant systemic infection by tobacco mosaic virus (Chen and Citovsky, 2003) and maintenance of the tomato fruit tissue integrity during post-harvest shelf life (Tieman and Handa, 1994). However, the actual physiological role of PME is still matter of controversy.

Four papers to our knowledge had been published in which the cloning of regulatory sequences of PME from higher plants was described. However, in all of them DNA comparison by computer was the only tool used to prove that the gene located downstream was indeed encoding a pectin methylesterase. Albani *et al.* (1991) reported the finding of a genomic clone from *Brassica napus* which contains the PME gene and its 5′ upstream regulatory region. Studies were conducted using a piece of the gene located dowstream as a probe. This gene was found to be expressed mainly during pollen development. Two putative PME promoter regions were cloned from *Brassica campestris* (Kim *et al.* 1997). Study of their sequence found them to have high homology with the previously reported promoter

PME from *Brasicca napus* (Albani *et al.* 1991). Further, a sequence motif similar to the one known to exist in two tomato pollen-specific promoter was located.

Tobacco transgenic plants with constructs containing two different promoter sizes from one of those two promoter available were made. Expression of the GUS gene was only detected during developing and mature pollen grains germinated in vitro. Recently, the cloning of two 5′ upstream region of PME from *Citrus sinensis* was published. Northern blot analysis showed that both DNA regulatory region are active in most of vegetative tissues (Nairn *et al.* 1998).

In our laboratory, we have cloned a 13.7 kb. genomic DNA from tomato containing a 2.59 kb. of DNA 5′ flanking region, along with all the PME genomic clone.

Identification of the protein encoded by the gene downstream was made by creating tobacco transgenic plants over expressing the PME cDNA (Gaffe *et al.* 1997) and comparing the sequences of the genomic and cDNA regions. In this work, we describe the study of a 5′ flanking region of a PME gene called pmeu1 (which stands for PME ubiquitous one) using computer tools and tobacco transgenic plants.

Cloning of the Genomic Fragment

The plaque lift technique was used to screen 810,000 clones of a tomato cv 'Cherry' genomic library made in EMBL3A ë phage with a radiolabeled small piece of PME; cloned by RT-PCR from tomato roots poly-A$^+$ mRNA (Gaffe *et al.* 1995). We found several hybridising, putatively positives, plaques. From every plate, we made two lifts and only that plaques producing signal in both lifts were chose to continue. After four rounds of purification and screening, one plaque turns out to be positive. Elimination of the bacteria present in the agar was made by using chloroform-containing SM buffer. For phage amplification, *E. coli* strain LE 392 was infected with the phage after cultured in LB media. DNA isolation from the phage was made through a phenol chloroform protocol (Ausubel *et al.* 1988).

Digested DNA with several restriction enzyme was separated by electrophoresis and blotted into nylon membranes. These DNA blots were probed with the PME cDNA complete sequence available to locate the phage DNA region encoding the genomic PME gene and the region 5′ upstream. Sal I and EcoR I digested DNA fragments were subcloned into pBSKS (+/-) vector (Stratagene Cloning Systems. La Jolla, CA).

The EMBL3A e phage library screened was created using the Sal I restriction site. Because of this, digested DNA fragments using Sal I were used to calculate the size of the tomato DNA inserted into the phage isolated, found to be 13.7 kb. All the procedure above mentioned was performed essentially as described (Sambrook *et al.* 1989) unless otherwise indicated.

DNA Sequencing of the Promoter and the Genomic Coding Region

Nested unidirectional deletions of the 5′ upstream DNA Sal I fragment were made by following the recommendations of the company (Erase-a-Base System, Promega Corporation, Madison, WI). Deleted clones with about 250 bp of size difference were used for DNA sequencing using the T_3 universal primer by the Sanger dideoxy chain termination technique following the recommendations (Sequenase Kit, United States Biochemical Corporation, Cleveland, Ohio). Second strand sequencing was determined by the DNA sequencing facility of IOWA State University by using primers designed at proper positions in the sequence (Iowa State University, Ames, Iowa).

Creation of Constructs

Three chimeric constructs driving the a-glucuronidase gene (*uidA*) under different sizes (2.596 Kb, 1.306 Kb and 0.267 Kb) of the promoter region were created by transcription fusion through the insertion of two stop codons in between the ATG of the pmeul gene and the ATG of the *uidA* gene. Every chimeric construct was ligated into the promoterless binary vector pBI101.3 (Bevan, 1984). This plasmid includes the neomycin phosphotransferase gene (NPTII) which confers resistance to kanamycin to be used as selectable marker. Furthermore, in this plasmid the DNA introduced is located between the right and left borders of the T-DNA, which allows the transference into the plant genome by *Agrobacterium* infection (Hooykaas, 1989; Zupan and Zambryski, 1995; Nester *et al.* 1996). Proper insertion of the different promoter sequences into the plasmid was confirmed by DNA digestion using suitable restriction enzymes and PCR using primers designed against sequences in the PME promoter and *uidA* gene.

Every chimeric construct created included 150 bp in between the ATG of the pmeu1 gene and the ATG of the *uidA* gene, containing sequences from the pmeu1 gene and pBSKS(+/-) phagemid and pBI101.3 binary vector. Sequencing between the two ATG's was used to verify the presence of two stop codons and to corroborate the transcription fusion of the two ATG.

Tobacco Transgenic Plants

Mobilisation of the pBI101.3 plasmid into *Agrobacterium* LBA4404 was performed by triparental mating using the broad-host helper plasmid pRK2013 (Ditta, 1981). *Agrobacterium* transconjugants were screened on plates containing a mixture of kanamycin and rifampicin antibiotics in YEP media (Sambrook *et al.* 1989). Verification of the mobilisation of the constructs was made by purification DNA from *Agrobacterium* following the recomendations (Wizard Minipreps, Promega Corporation, Madison, WI) and digestion with proper restriction enzymes. Tobacco (*Nicotiana tabaccum* W38) young leaves and tomato (*Lycopersicon esculentum* cv Ohio 824) cotyledons were infected with *Agrobacterium* by using the leaf disk technique (Mathis and Hichee, 1994) and selection for transformants was done by using kanamycin in the media.

GUS Activity Measurement

After induction of roots, about 50 primary independent transformant plants growing *in vitro* harbouring every of the three constructs were selected at random to measure GUS activity in leaf. This was done using the fluorometric technique (Jefferson *et al.* 1987) with a Perkin Elmer LS5 fluorometer. Quantification of reaction product was done by using a 4-methylumbelliferone standard curve. Also, six independent transgenic plants were used to measure GUS in root, stem and leaf tissues. Every GUS measurement was done at least three times. For enzymatic specific activity, protein determination was made using Bradford (1976) with bovine serum albumin as standard.

In order to examin the GUS presence in different tissues, at least 20 primary transgenic plants harbouring the different constructs, were vacuum infiltrated with a 1.9 μM solution of 5-Bromo-4-Chloro-3-Indolyl-Glucuronide (Jefferson *et al.* 1987) as described (Mandel *et al.* 1995).

Pollen Germination *In Vitro*

Tobacco flowers in anthesis were collected from plants growing in the greenhouse and transported immediately to the laboratory. Anthers were cut and only that pollen released by a gently shaking was used for germination studies. Pollen was germinated using the Brewbaker and Kwack solution as described (Brewbaker and Kwack, 1963). Histochemical GUS staining was performed after four hours of pollen germination. Germination solution was changed by the GUS staining solution and left at 37°C for at least 18 hrs before examination for GUS staining.

Computer Analysis

DNA sequence from the different *pmeu1* 5′ flanking region deletions were joined together using DNAsis (Hitachi Software Engineering Co., LTD., 1991). Comparison between the *pmeu1* cDNA and *pmeu1* genomic clone was performed using Harr plot analysis with DNAsis software. Presence of known cis-acting elements was determined using the programmes MathInspector ver. 2.2 (Quandt *et al.* 1995), TFSEARCH ver. 1.3 (Parallel Application, Tsukuba Laboratory, RWCP, Japan),

Signal Scan ver. 4.05 (Prestridge, 1991) and Pattern Search (Wingender *et al.* 1996; Wingender *et al.* 1997).

Per cent of identity among the different PME promoters and PME transcribed regions were determined using Align (Myers and Miller, 1988). Alignment of deduced amino acid sequences was performed using GCG's Pileup Programme (Genetics Computer Group, Madison, WI).

Multiple sequence alignment was performed using CLUSTAL W (Thompson *et al.* 1994). DNA direct repeats for the tomato PMEU1 promoter were determined using Proscan ver 1.7 and repeats from GCG software ver. 9.0 (Genetics Computer Group, 1995).

Perfect inverted repeats (mirror repeats) were located using Palindrome from GCG software ver 9.0 (Genetics Computer Group, 1995). Putative TATA box was located by Signal Scan ver. 4.05. Phylogenetic analysis were done using the phylogeny inference package (Felsenstein, 1989; Felsenstein, 1993).

Statistical Analysis

Comparison of leaf GUS activities for the three constructs and for the different tissues was made by variance analysis using a completely randomised design for unbalanced number of repetitions. Tukey test was used when needed to find differences among means. Because it is known that the GUS enzymatic activity in populations of first-generation transgenic plants does not follow a normal distribution (Nap *et al.* 1993), we performed a Box-Cox transformation before variance analysis.

Biodegradation of Natural Phenolic Compounds as Single and Mixed Substrates by Fusarium Flocciferum

Environmental pollution due to the release of natural phenolic compounds from agroindustrial operations has become widespread in the world. The structure of the compounds present is similar in many industrial effluents and residues like those produced in wine-distillery, olive oil extraction, green olive debittering, cork preparation, wood debarking and coffee production (Field and Lettinga, 1991; Borja *et al.* 1993; Sayidi and Ellouz, 1995; Brand *et al.* 2000; Lesage-Meessen *et al.* 2001; Minhalma and de Pinho, 2001; Aggelis *et al.* 2002).

The biological treatment of industrial wastewaters usually depends upon the oxidative activities of microorganisms. Filamentous fungi can be an important source of phenol-degrading species as they grow frequently in wood where phenolic structures are present. Nevertheless filamentous fungi are not frequently used due to difficulties in their cultivation in liquid media and their slow growth rate in comparison with most of the other microbial species (Righelato, 1975). Moreover, organic compounds like phenol and its derivatives have toxic effects that limit the biological treatment, because they can be growth-rate inhibitory even to species that have the metabolic capability of using it as a substrate for growth (Evans, 1963).

The purpose of this study was to investigate the ability of the fungus *Fusarium flocciferum* to degrade some phenolic compounds namely, gallic, protocatechuic, vanillic, syringic, caffeic and ferulic acids and syringic aldehyde, usually present in agroindustrial effluents. The results of the chemical and ecotoxicological monitoring during biodegradation experiments are presented and discussed having in mind the development of a biotreatment capable of degrading these compounds and detoxifying these wastes.

Materials and Methods

Where k and n are Freundlich constants, which correlated to the maximum adsorption capacity and adsorption intensity, respectively.

Linearised Freundlich Equation

$$\text{Log } Q = \text{Log } k + 1/n \text{ Log } C_f$$

In the biosorption of the tested metals by the different *Pseudomonas* species, most of the metal ions were sequestered very fast from solutions within the first 10 minutes and almost no increase in the level of bound metals have been occurred after this time interval (data not shown).

The comparison of the sorption performance of the different biosorbents was achieved under the same environmental conditions (*i.e.* pH, temperature, agitation speed, etc.). Biosorption equilibrium isotherms were plotted for the metal uptake Q against the residual metal concentrations in solution. The Q versus C_f sorption isotherm relationship was mathematically expressed by linearised Langmuir and Freundlich models. The higher the values of k and n and the lower the value of b, the higher the affinity of the biomass (Asku *et al.* 1991; Jalaliet al. 2002).

Quantitative Real-time PCR Method to Detect Changes in Specific Transcript and Total RNA Amounts

In tissue exposed to a given treatment, the level of expression of several genes may change significantly, possibly altering the total amount of RNA in the tissue. In studies to determine the change in the level of expression of a given gene of interest, methods such as Northern blots, microarray interrogation, or quantitative real-time PCR typically are used to estimate the number of copies of a transcript of interest in an aliquot of total RNA extracted from tissue samples before and after treatment (Xu *et al.* 2000; Maguireet *et al.* 2002). This determination is influenced by the amount of total RNA in the tissue. For example, if total RNA increased 2-fold, and the occurrence of the target transcript increased 10-fold, estimating the number of transcript copies in an aliquot of RNA would indicate only a fivefold increase in the target transcript, while a 10-fold increase had actually occurred in each cell. Here, we suggest a simple method using qRT-PCR to estimate the relative proportions of total RNA and a specific RNA transcript in two tissue samples.

Near isogenic wheat (*Triticum aestivum* L.) lines 442 and 443, differing only in the region of chromosome 5A carrying gene(s) conditioning winter wheat vs. spring wheat growth habit (Storlie *et al.* 1998) were used. Line 443 exhibits spring wheat growth habit and 442 exhibits winter wheat growth habit. Plants were grown under 16 hrs light at 20°C for 14d, then were transferred to a chamber with the same light conditions at 2°C. The second leaves (of three) to open on the plants were collected at the time of transfer (control) and after 7d growth at 2°C (cold-exposed treatment).

To determine the amount of water in the plant samples, second leaves from each of the treatments were weighed, dried for 24 hrs at 80°C, and then weighed again. Water content was determined by subtraction. All moisture determinations and PCR analyses were replicated three times using independent plants.

Three types of samples were processed; (i) tissue from control plant; (ii) tissue from cold-exposed plant; (iii) a sample consisting of equal amounts of tissue from control and treated plants. The same mass of tissue from each plant was pulverised in a mortar and pestle in liquid nitrogen.

Trizol reagent (Invitrogen, San Diego, CA) was added to the mortar and total RNA was extracted according to the manufacturers directions. The final RNA pellet was dissolved in DEPC-treated water and quantified with UV spectrophotometry. Quantitative real-time PCR was carried out on a RotorGene 2000 (Corbett Research, Sydney, Australia) using a SYBR green detection protocol (Karsai *et al.* 2002). Synthesis of cDNA and PCR amplification was carried out with the Superscript One-step RT-PCR Platinum Taq kit (Invitrogen). The amplification protocol consisted of two steps, cDNA construction for 15 min at 50°C followed by a PCR profile of a 3 min denaturation at 95°C, then 32 cycles of 15 s at 95°C, 20 s at 54°C, and 17 s at 72°C. The amount of product was determined at the end of each cycle by the RotorGene software. The reaction mixture provided in the Superscript kit was adjusted to 2mM $MgCl_2$, 100nM primers, 1:40,000 SYBR Green (Molecular Probes, Eugene, OR), and 125 ng total RNA in 20 μl reactions. A cloned wheat MnSOD gene was used as template in amplification reactions to develop a standard curve for quantification of the products of MnSOD and PLD from each replication.

The template for qRT-PCR consisted of six mixtures of RNA from the control and cold-exposed plants. A seventh preparation was made up of RNA extracted from combined equal amounts of plant tissue from the two temperature regimes.

Recovered RNA had clearly distinguishable rRNAs, indicating large RNA fragments had been recovered. Regression of copy number on the proportion of RNA from the cold-treated plant indicated that combining RNA from the control and cold-exposed plants resulted in a linear, additive, determination of copy number. Therefore, if the amounts of total RNA per unit mass of tissue were the same in the cold-exposed and control plants, the number of copies of transcript extracted from the tissue sample comprised of equal masses of the two plants would be equal to the number of copies of transcript observed in the 0.5:0.5 mixture of RNAs extracted from the two plants individually.

However, the qRT-PCR result from the RNA extracted from the combined plant tissues indicated a value significantly greater than the 0.5:0.5 mixtures with both MnSOD and PLD. Therefore, the plant with the higher number of copies of transcript per nanogram RNA, the cold-treated plants in this case, also contributed more RNA to the reaction template from the combined tissues, *i.e.* contained more RNA per unit mass of tissue. The ratio of the amount of RNA in the cold-treated to control plants may be calculated from the qRT-PCR results from the control, cold treated, and mixed tissue samples as follows.

In one replication of the MnSOD determination from wheat line 443, the number of copies of MnSOD transcript in 125ng RNA from the control plant was 339875; from the cold-treated plant the number was 1297629; and from the RNA extracted from the combined plant tissues the number was 983264. From this information the proportion of the total RNA extracted from the combined tissue samples that originated from the control plant, and the proportion that originated from the cold-treated plant, can be calculated as:

Let: T=nanograms RNA from the cold-treated plant and C=nanograms RNA from the control plant. The average number of copies of MnSOD transcript per nanogram of RNA from the control plant was 339875/125=2719, the average number of copies of transcript per nanogram of RNA from the cold-treated plant was 1297629/125=10381, and the number of copies of transcript in 125ng RNA extracted from the combined tissues was 983264. Thus: (Cng)(2719 copies/ng) + (Tng)(10381 copies/ng) = 983264 copies, or: 2719C + 10381T=983264 copies (equation 1) and: T+C = 125ng (equation 2). From equation (2), T=125-C, thus, equation (1) can be rewritten: 2719 C+ 10381 (125-C) = 983264, which is solved for C=41.0, leading to: T = 125-C = 125-41.0 = 84.0. Therefore, in the combined tissue sample, 2.05 times (84/41) the amount of total RNA originated from the cold-treated plant as compared to the control plant.

This calculation was repeated for each of three replications. These results were compared to the result of the linear regression equation developed from mixing RNAs from the control and from the cold-exposed wheat plants. The mean copy numbers of MnSOD from the total RNA of mixed plant tissue, corresponded to a proportion of 0.697 RNA originating from the cold-treated plant.

Hence, 1-0.697=0.303 originated from the control plants, indicating a 0.697/0.303 = 2.30 fold increase of total RNA in the cold-exposed plants, identical to the mean of the results of three replications of the calculations demonstrated above, suggesting the simpler, calculation-based method provides a reliable estimate of the total RNA fold change.

Comparison of the average number of MnSOD transcript

copies per nanogram RNA from the control and cold-treated plants indicated a 3.5-fold increase in the cold-treated over the control. Taken together, these two results indicated that the cold treatment resulted in an increase in total RNA per unit mass of plant tissue of 2.3-fold, and a 3.5-fold increase in the measured concentration of MnSOD transcript per unit mass of RNA, resulting in a total increase of (2.3)(3.5) = 8.1-fold increase of MnSOD transcripts per unit mass of plant tissue.

Similar calculations with the PLD transcript copy number data also indicated a 2.3 fold increase of total RNA per unit mass of tissue, and a 2.7 fold increase of PLD transcript per unit mass of RNA, suggesting in a 6.2 fold increase of PLD expression per unit mass of tissue in response to exposure to cold.

Others have reported that the expression of MnSOD increases more rapidly in spring wheat, compared to winter wheat, after exposure to cold temperature (Wu *et al.* 1999). A comparison of the fold increase of MnSOD expression in 442 and 443 supports that observation, and suggests that most of the relative increase seen in 443, compared to 442, was due to an increase in the number of copies per unit mass of RNA, with about the same amount of increase of total RNA per unit mass of tissue seen in 442 and 443.

Determinations of water content revealed that the plants exposed to cold temperatures contained an average of 4 per cent less water than the control plants. This difference was statistically significant, but there was no significant difference between plant sources (statistical separation not shown).

Assuming that this 4 per cent difference in water content resulted in a 4 per cent increase in the number of cells per unit mass of tissue, the samples consisting of equal masses of plant tissue from the two temperature treatments would actually contain more cells of the cold treated tissue than the control tissue, resulting in an inflated estimate of RNA content in the cold-treated tissue.

Using qRT-PCR method, we calculated an average increase

of total RNA, including the spring and winter wheat lines, of 2.3-fold (1.9-fold when corrected for change in water content) after seven days of cold acclimation. To our knowledge, this method of using qRT-PCR with RNA extracted from equal masses of control and treated tissues is the first attempt to quantify changes in the amount of total RNA with qRT-PCR. Sarhan and D'Aoust, 1975, using a method of extraction and quantification of nucleic acids, first measured RNA change after 11 days of cold acclimation and estimated an average increase of total RNA, including spring and winter wheat, of 2.1-fold; water content apparently was not measured in that study. Our finding of an average 2.3-fold increase in total RNA and the 2.1 fold increase found by Sarhan and D'Aoust, 1975 were within one standard deviation unit of our data, and therefore, were not considered significantly different, suggesting the qRT-PCR method we propose here provides a rapid and reliable method of estimating changes in total RNA content in response to treatments.

Copper Removal from Aqueous Solution by Marine Green Alga Ulva reticulate

Heavy metals released into the environment by technological activities tend to persist indefinitely, circulating and eventually accumulating throughout the food chain, becoming a serious threat to the environment. The main sources of heavy metal pollution include electroplating, painting, dying, surface treatment industry, etc. These pollutants are toxic and non-biodegradable and probably have health effects. Conventional processes for heavy metal removal often are neither effective nor economical (Volesky and Holan, 1995). Biosorption of heavy metals is one of the most promising technologies involved in removal of toxic metals from industrial waste streams and natural waters (Pagnanelli *et al.* 2000). Various biomaterials have been examined for their biosorptive properties and different types of biomass have shown levels of metal uptake high enough to warrant further research (Volesky and Holan, 1995). Among the most promising types of biosorbents studied is algal biomass (Figueira *et al.* 2000). Marine algae proliferate ubiquitously and abundantly in the littoral zones of world oceans, they are rather stable and fast growing (Yang and Volesky, 1999).

Most of the metal uptake studies are focused on brown marine algae (Figueira *et al.* 2000; Cossich *et al.* 2002), due to its extreme metal adsorbing potential. Among the few studies on green algae, Darnall *et al.* 1986 inferred that *Chlorella vulgaris* has high affinity for the binding of Au^{3+}, Ag^{+}, and Hg^{2+} ions, whereas Holan and Volesky, 1994 inferred that among the algal strains examined, *Codium taylori* showed least affinity for nickel. *Ulva reticulata* is a very common green alga found throughout the Palk bay, and its potential utilisation as a metal biosorbent is of interest.

Copper, one of the most widely used heavy metals, is mainly employed in electrical and electroplating industries, and in larger amounts is extremely toxic to living organisms. The presence of copper (II) ions cause serious toxicological concerns, it is usually known to deposit in brain, skin, liver, pancreas and myocardium (Davis *et al.* 2000). The objective of the present work was to test the ability of *Ulva reticulata* to remove copper (II) ions from aqueous solution. Also basic parameters of copper biosorption equilibrium were determined at various pH values. Copper desorption was also examined.

Ulva reticulata was collected and sun dried in Mandapam (India). Dry biomass was cut with a knife into irregular shaped particles between 1 and 3 mm in size. The biomass was then washed with distilled water, then filtered and finally dried overnight at 60°C and subsequently used for sorption experiments.

Analytical grades of $CuSO_4.5H_2O$, HCl, H_2SO_4, HNO_3, NaOH, and $CaCl_2.2H_2O$ were purchased from Ranbaxy Fine Chemicals Ltd., India. Cupric (Cu^{2+}) ions were prepared by dissolving its corresponding sulphate salt in distilled water.

Biosorption experiments were performed in a rotary shaker

at 150 rpm using 250 ml Erlenmeyer flasks containing 100 ml of different copper concentrations. After 6 hrs contact (according to the preliminary sorption dynamics tests), with 0.2 g *U. reticulata* biomass, equilibrium was reached and the reaction mixture was centrifuged at 3000 rpm for 10 min.

The metal content in the supernatant was determined using Atomic Absorption Spectrophotometer (AAS 6VARIO; Analytik Jena, Germany). The amount of metal adsorbed by alga was calculated from the differences between metal quantity added to the biomass and metal content of the supernatant using the following equation:

$$Q = (C_0 - C_f) * V / M \quad (1)$$

where Q is the metal uptake (mg/g); C_0 and C_f are the initial and equilibrium metal concentrations in the solution (mg/L), respectively; V is the solution volume (L); and M is the mass of biosorbent (g). The pH of the solution was adjusted by using 0.1M HCl or 0.1M NaOH.

The Langmuir sorption model was chosen for the estimation of maximum copper sorption by the biosorbent. The Langmuir isotherm can be expressed as:

$$Q = \frac{Q_{max} b C_1}{1 + b C_1} \quad (2)$$

where Q_{max} is the maximum metal uptake (mg/g) and b is the Langmuir equilibrium constant (L/mg). For fitting the experimental data, the Langmuir model was linearised as follows:

$$1/Q = 1/ Q_{max} + 1/ b\, Q_{max}\, C_f \quad (3)$$

The Freundlich model is represented by the equation:

$$Q = KC_f^{1/n} \quad (4)$$

where K (L/g) and n are constants. For fitting the experimental data, the Freundlich model was linearised as follows:

$$\ln Q = \ln K + 1/n \ln C_f \quad (5)$$

The metal-loaded biomass after biosorption was contacted with elutants, 0.1 M HCl, 0.1 M H_2SO_4, 0.1 M HNO_3 and 0.1 M $CaCl_2$ in 250 ml Erlenmeyer flasks for 3 hrs on a rotary shaker (150 rpm) to study the removal of biosorbed metal ions. The remaining procedure was same as that in the sorption equilibrium experiments. After desorption, the biomass was washed with distilled water, then filtered and finally dried overnight at 60°C. The loss in biomass weight was calculated and the biomass was subsequently used for resorption studies.

Effect of pH on Copper Biosorption

The initial pH of the solution is a very important factor in copper sorption uptake by *U. reticulata*. In general very little or no copper uptake was observed at pH less than 2.0. Working over pH 6.0 was avoided to prevent the possible precipitation of copper hydroxide. Little copper uptake at low pH values is an indication of competition of excess of protons for the same binding sites on the algal cell wall (Puranik and Paknikar, 1999). The copper uptake increased with increasing pH, to the maximum near pH 5.5, then decreased at higher pH value.

Langmuir and Freundlich sorption models were used to evaluate the sorption behaviour of the biosorbent. Langmuir sorption model served to estimate the maximum metal uptake values where they could not be reached in the experiments. The constant b represents affinity between the sorbent and sorbate. The Langmuir model parameters were largely dependent on the initial solution pH values. Both the maximum metal uptake Q_{max} and the Langmuir equilibrium constant b increases with increasing initial pH from 3 to 5.5 and decreased, when the pH was increased further. It is worth noting both K and n values also reached their maximum values at pH 5.5, this implies that the binding capacity reaches the highest value and the affinity between the biomass and copper ions was also higher than at other pHs investigated. At all initial pH values investigated (except at pH 4.5), Freundlich model resulted in higher correlation coefficients than those obtained for Langmuir model fits.

Kinetic Studies

On changing the initial copper concentration from 250 to 1000 mg/L, the amount sorbed increased from 36.71 to 70.28 mg/g at pH 5.5. But the per cent removal of copper decreased from 29.37 to 14.06 per cent as the copper concentration increases from 250 to 1000 mg/L. This is because at lower concentration, the ratio of the initial moles of copper (II) to the available surface area is low and subsequently the fractional sorption becomes independent of initial concentration. However, at higher concentration the available sites of sorption becomes fewer compared to the moles of copper (II) present and hence the percentage removal of copper (II) is dependent upon the initial Cu^{2+} concentration. To analyse the sorption rates of copper (II) ions onto the biomass, two simple kinetic models were tested.

Pseudo-first-order model. The Pseudo-first order rate expression is generally described by the following equation (Ho and McKay, 1998),

$$dQ_t/dt = K_1\ (Q_e - Q_t) \tag{6}$$

where, Q_e is the amount of copper sorbed at equilibrium per unit weight of sorbent (mg/g); Q_t is the amount of copper sorbed at any time (mg/g); and K_1 is the rate constant (min^{-1}).

Integrating and applying boundary conditions, t = 0 and $Q_t = 0$ to t = t and $Q_t = Q_t$, equation (6) takes the form:

$$\log\ (Q_e - Q_t) = \log\ Q_e - K_1\ t\ /\ 2.303 \tag{7}$$

In order to obtain the rate constants, the straight-line plots of log $(Q_e - Q_t)$ against t were made for *U. reticulata* at different initial copper concentrations. The intercept of the above plot should equal to log Q_e. However, if the intercept does not equal to the equilibrium copper uptake then the reaction is not likely to be first order even this plot has high correlation coefficient with the experimental data (Ho and McKay, 1998). Correlation coefficients were found to be between 0.8349 and 0.9519, but the calculated Q_e is not equal to experimental Q_e, suggesting the insufficiency of Pseudo-first-order model to fit the kinetic data for the initial copper concentrations examined. Pseudo-second-order model. The sorption data were also analysed in terms of a Pseudo-second order mechanism (Ho and McKay, 1998), given by:

$$dQ_t/dt = K_2\ (Q_e\text{-}Q_t)^2 \qquad (8)$$

where K_2 is the equilibrium rate constant (g/mg min). Integrating the above equation and applying boundary conditions t = 0 and $Q_t = 0$ to t = t and $Q_t = Q_t$, gives:

$$t/Q_t = 1/h + 1/Q_e \qquad (9)$$

where *h* is the initial sorption rate. If second order kinetics is applicable, the plot of t/Q_t against *t* of equation (9) should give a linear relationship from which the constants Q_e, *h* and K_2 can be determined.

Good fits were observed for all initial concentrations indicating that sorption reaction can be approximated with the Pseudo-second-order model. Correlation coefficients were found to be between 0.9999 and 1.0000. The lowest correlation coefficient in this case was 0.9999, which is even better than the first order model. The values of equilibrium sorption capacity, Q_e, were found to increase from 37.04 to 71.43 mg/g, for an increase in the initial concentration from 250 to 1000 mg/L, which also shows good agreement with the experimental equilibrium copper uptake. The very fast sorption kinetics observed with the *U. reticulata* biomass represents an advantageous aspect when effluent treatment systems are designed.

Desorption of Copper

The copper-loaded biomass was eluted by various elutants, 0.1 M $CaCl_2$, 0.1 M HCl, 0.1 M H_2SO_4 and 0.1 M HNO_3. The optimal elutant must be effective, non-damaging to the biomass, non-polluting and cheap. In this work, the effect of 0.1 M $CaCl_2$ on metal elution was examined at different initial pH conditions. Elution efficiency (E) was determined by the ratio of the metal mass in the solution after desorption to the metal mass initially bound to the biomass (Davis *et al.* 2000).

The elution efficiency of 0.1 M $CaCl_2$ was strictly pH dependent and maximum desorption performance was observed at initial pH 3. The biomass that was exposed to $CaCl_2$ has remained soft and appears to be suitable for subsequent cycle. The mineral acids wash of copper-laden biomass released all the metal ions.

However, the biomass become more fragile and lost its original colour. The biomass weight loss was greater than 35 per cent, when mineral acids are used as elutants. Acid solutions could dissolve some types of polysaccharides that may contain ion-binding sites and the mineral contents of the biomass as well (Kuyucak and Volesky, 1989). Several researchers also observed similar deterioration in the macroscopic appearance of the biosorbent material. For instance, Davis *et al.* 2000 observed *Sargassum filipendula* biomass become more fragile and the colour of the biomass turned from brown to black, when exposed to stronger HCl solution.

An important parameter for metal biosorption is the Solid-to-Liquid ratio (S/L) defined as the mass of metal-laden biosorbent to the volume of the elutant (Davis *et al.* 2000). Upon elution of the metal from the biosorbent, it is desirable to use the smallest possible eluting volume so as to contain the highest concentration of the metal. At the same time, the volume of the solution should be enough to provide maximum solubility for the metal desorbed.

The elution efficiency of 0.1 M $CaCl_2$ decreased with increasing S/L ratio at all the pH values examined. For instance, 0.1 M $CaCl_2$ (in HCl, pH 3), at S/L 1 g/L elution efficiency was 90.25 per cent and dropped to 52.88 per cent at S/L 10 g/L. Among the mineral acids examined, the elution efficiency of 0.1 M HCl appeared to be nearly independent of the S/L ratio up to the examined value of 10 g/L. Elution efficiency of 0.1 M H_2SO_4 and 0.1 M HNO_3 dropped to 92.71 per cent and 95.04 per cent respectively at S/L 10 g/L. It was clear that protons played a major role in desorption. The increase in final pH observed in desorption experiments indicate that copper was displaced from the biomass by protons, showing the probable involvement of ion exchange.

The use of *U. reticulata* as a potential biosorbent depends not only on the biosorptive capacity, but also on how well the biomass can be reused. The biomass was used for three sorption-desorption cycles and regenerated using 0.1 M $CaCl_2$ (in HCl, pH 3) at S/L 1 g/L. At the end of three cycles the sorption capacity of regenerated *U. reticulata* decreased by 13.2 per cent and the weight loss was 15 per cent. The elution efficiency of 0.1 M $CaCl_2$ (in HCl, pH 3) almost remained constant throughout the three cycles examined. The present experiments revealed 0.1 M $CaCl_2$ (in HCl, pH 3), appears as an appropriate metal-eluting agent for *U. reticulata*, with its low biomass damage adequate to be employed in continuous operation.

The present work evaluated the removal of copper (II) from aqueous solution using *U. reticulata* biomass. The sorption capacity of copper on *U. reticulata* increased with increase in pH reaching a maximum at 5.5. The equilibrium sorption data agrees well with Freundlich isotherm with high correlation coefficients. The kinetics of the sorption process was found follow Pseudo-second-order model.

The highest copper elution efficiency was 100 per cent for all mineral acids examined at S/L 1g/L. $CaCl_2$ (in HCl) exhibited a maximum elution efficiency of 90.25 per cent at pH 3 and S/L 1g/L, and decreased as the S/L increases. Low biomass damage and high elution efficiency in the subsequent sorption-desorption cycles favour $CaCl_2$ (in HCl, pH 3) to be employed as a suitable metal desorbing agent for *U. reticulata*.

Isolation of *adh* mutant of *Lactobacillus rhamnosus* for production of L(+) Lactic acid. L(+) lactic acid production is gaining interest due to its applications in food and pharmaceutical industries (Harrig, 1989; Datta, 1995). Most homofermentative organisms produce both the stereoisomeric forms of lactic acid. Heterofermentative organisms may produce lactic acid along with other metabolites like ethanol, acetate, formate, etc., anaerobically. The purification of one form of lactic acid from its racemic mixture is very difficult, which may involve expensive chromatography techniques. The most suitable way is to select the L(+) producer organism or use genetically modified strains. A few attempts have been made to improve and modify the production of L(+) lactic acid by metabolic engineering in *Lactobacillus* species. The *ldh*D gene in *L. helveticus* was inactivated (Bhowmik *et al.* 1993). A twofold increase in the amount of L(+) lactic acid was observed which restored the amount of total lactic acid to the level in the wild type strain. The *ldh*L gene was overexpressed in *L. plantarum* by increasing its copy number (Ferain *et al.* 1994). This increased the activity of L-lactate dehydrogenase (L-LDH) but was found ineffective for L(+) lactic acid production. The total lactic acid production was not affected by the inactivation of *ldh*D gene or inactivating the *ldh*L gene.

In another study, two *ldh*D$^-$ strains of *L. helveticus* were constructed by deleting the promoter region of the *ldh*Dgene in the first construct and by replacing the structural gene of *ldh*D gene by *ldh*L gene of the same species in the second construct (Nikkila *et al.* 2000). This has not only increased the enzyme activity of L-LDH by 53 per cent and 93 per cent respectively, but also enhanced L(+) lactic acid production. However, they could only produce the lactic acid rich in L(+) form, but there was a need to separate L(+) lactic acid from its D(-) form to facilitate its usage for medical field. The presence of D(-) lactic acid is interfering in the L(+) lactic acid polymer formation which is applied for medical purposes. In *Lactococcus lactis*, *ldh*Lgene is located as a part of the *lac* operon. An increase in copy number of the plasmid containing the whole operon resulted in a slight increase in lactic acid production (Llanos *et al.* 1992). The *ldh*D gene of *L. johnsonii* was altered by an 8-bp deletion. The D-LDH enzyme activity was completely absent in the mutant strain and the remaining L-LDH enzyme activity rerouted the carbon flux to L-lactate with a marginal increase in the secondary end products (Lapierre *et al.* 1999).

Many of the genetically modified homofermentative lactic acid bacteria were unable to produce pure L(+) form of lactic acid. The present studies were planned to produce L(+) lactic acid as the only product using the heterofermentative organism *Lactobacillus rhamnosus*. This organism produces ethanol along with L(+) lactic acid under anaerobic conditions. For this purpose *adh*⁻ mutant of *L. rhamnosus* was developed by chemical mutagenesis. Its behaviour was examined in batch fermentation. The advantage of this method is the pure L(+) form of lactic acid production as the sole product. The downstream processing for the separation of pure L(+) lactic acid from the fermentation broth can be made easier if a defined media is used. The concentration of amino acids, vitamins and trace metals solution were optimised by experimental design and response surface method and their batch kinetics were observed in the optimised media.

Microorganism and Culture Condition

The bacterial strain *Lactobacillus rhamnosus* MTCC 1408 (obtained from Institute of Microbial Technology, Chandigarh) is used. Stock cultures of the bacteria were stored in MRS medium as 15 per cent glycerol stock at -80°C (as instructed by Institute of Microbial Technology, Chandigarh). The medium used in the 2 per cent (w/v) agar plates, precultures, and batch fermentations contained 100 g/l glucose, 30 g/l yeast extract, 0.2 g/l KH_2PO_4, 0.2 g/l K_2HPO_4, 0.3 g/l $MgSO_4.7H_2O$, 0.03 g/l $MnSO_4.H_2O$, 0.03 g/l $FeSO_4$, 2 g/l Sodium succinate, 2 g/l Sodium sulphate. The pH of the medium was adjusted at 6.2 by adding 2N NaOH/ 2N HCl before sterilisation (Srivastava *et al.* 1992). *L. rhamnosus* was anaerobically grown on 2 per cent (w/v) agar plates at 40°C for 24 hrs, and then kept at 4°C. Initial preculture for bioreactor cultivations was grown in glucose-yeast extract medium at temperature 40°C for 24 hrs inside anaerobic jars. 2 per cent (v/v) inoculum was used from this culture to prepare the successive preculture. This was further used for inoculating the fermentor.

Bioreactor Cultivations

Bioreactor cultivations were performed in a 3.0 l glass fermentor (KLF 2000, Bioengineering AG, Wald Switzerland) using 2 l working volume. The fermentor was equipped with all monitors and controllers for pH, temperature and dissolved oxygen. An agitation speed of 300 rpm was maintained with the help of agitator fitted with two flat blade impellers. Anaerobic conditions were maintained by sparging nitrogen gas before inoculation and intermittently during the course of fermentation. The pH was maintained at 6.2 by automatic addition of 4N NaOH/ 4N H_2SO_4 during lactic acid fermentation.

Mutagenesis

Actively growing cells from the second preculture in glucose-yeast extract medium, were collected and centrifuged to separate cells from supernatant. The cells were then washed with 0.1 M citrate buffer (pH 5.5), resuspended in 0.1 M citrate buffer (pH 5.5), and then mutagenised by exposing these cells to N- methyl, N′-nitro, N-nitrosoguanidine (MNNG, 50 mg/ml) for 30 minutes at 28°C. Thereafter, cells were washed with 0.1 M phosphate buffer (pH 7.0). N- methyl, N′-nitro, N-nitrosoguanidine, breaks down at pH 7.0.

Selection of Mutants

The washed cells were suspended in glucose-yeast extract medium and spread onto agar plates 0.6 per cent (v/v) allyl alcohol. This compound inhibits wild type cells since cells with functional alcohol dehydrogenase activity convert allyl alcohol to the toxic compound acrolein (Pulci *et al.* 1993).

$$\underset{\text{Allyl alcohol}}{CH_2{=}CH{-}CH_2OH} \xrightarrow{\text{Alcohol dehydrogenase}} \underset{\text{acrolein}}{CH_2{=}CH{-}CHO}$$

Single colonies obtained on plates were repeatedly

transferred to agar plates containing 0.6 per cent (v/v) allyl alcohol to get a stable mutant.

Analytical Methods

Optical density of the culture was measured at 600 nm with a spectrophotometer (spekol 1100, analytik jena, Germany) The dry cell weight was determined by centrifuging the culture broth to separate cells from supernatant, then washing with 0.85 per cent NaCl solution and drying in preweighed aluminium foil cups at 70°C till constant weight. One optical density was found to be equivalent to 0.34 g dry cell weight per litre.

Concentrations of glucose in samples were determined by Dinitro Salycilic Acid method (Miller, 1959). Total lactic acid was determined colorimetrically using p-phenyl phenol (Barker and Summerson, 1941). L(+) lactic acid concentration were analysed spectrophotometrically at 340 nm using L-lactate dehydrogenase enzyme (L-LDH). The assay mixture containing 0.6 M glycine, 0.47 M hydrazine, 35 mg/ml NAD, 13.6mg/ml L-LDH and the appropriately diluted test samples were incubated at 30°C for 90 min before taking optical density at 340 nm.

The concentration of ethanol, acetate, formate were determined by a gas chromatograph (AIMIL, Nucon, India, Series 5765) equipped with Chromosorb 101 column using nitrogen as the carrier gas and a mixture of hydrogen and oxygen gas to sustain the flame. The detector, injector and oven temperature were maintained at 200°C, 195°C and 180°C respectively. A gas flow rate of 35 ml/min was maintained.

Optimisation of Media

A glucose limited defined media (composition obtained from Centre for Process Biotechnology, DTU, Denmark) was taken as the basis. Saline Solution I (containing NH_4Cl 0.142 g/l and K_2SO_4 0.059 g/l), Saline Solution II (containing KH_2PO_4 0.197 g/l, NaCl 3.247 g/l, $MgCl_2$ 0.118 g/l and $FeSO_4$ 3.09×10^{-3} g/l), trace metal solution (containing $(NH_4)_6(MO_7)_{24}$ 4.11x10^{-6} g/l, H_3BO_3 27.44 x10^{6} g/l,$CoCl_2$ 7.9 x10^{6} g/l, $CuSO_4$ 2.88 x10^{6} g/l,$MnCl_2$ 13.9 x10^{6} g/l, $ZnSO_4$ 3.1 x10^{6} g/l and$CaCl_2$ 8.17 x10^{6} g/l) and glucose (10 g/l) were sterilised separately at 121°C for 20 min. Amino acid solution (containing L-alanine 0.168 g/l, L-arginine 2.71 g/l, L-asparagine.133 g/l, L-cysteine 0.647 g/l, L-glutamate 0.163 g/l, L-glutamine 0.341 g/l, L-glycine 0.226 g/l, L-histidine 0.052 g/l, L-isoleucine 0.35 g/l, L-leucine 0.350 g/l, L-lysine 0.284 g/l, L-methionine 0.116 g/l, L-phenylalanine 0.220 g/l, L-proline 0.128 g/l, L-serine 1.354 g/l, L-threonine 0.224 g/l, L-tryptophan 0.227 g/l, L-tyrosine 0.06 g/l and L-valine 0.234 g/l) and vitamin solution (containing biotin 0.109x10^{-3} g/l, pyridoxal 2.28 x10^{-3} g/l, folic acid 1.128 x10^{-3} g/l, niacinamide1.083 x10^{-3} g/l, thiamine HCl 1.118 x10^{-3} g/l, calcium pantothenate 0.528 x10^{-3} g/l) were added by sterile filtration using 0.2 m filter. In this medium, 0.2 g/l of $MnSO_4$ was added as according to the composition of MRS medium composition recommended by Institute of Microbial Technology, Chandigarh. The experiments were carried out in 500 ml flasks with 200 ml medium, inoculated with 2 ml culture from preculture II, and incubated at 40°C for 48 hrs under anaerobic conditions without shaking.

Statistical Experimental Design

A Box Behnkehn experimental design with three variables, amino acids, vitamins, trace metals solution and the two parameters or responses, lactic acid and biomass were used which resulted in a total of 20 experiments. Design Expert version 5.5 (Statease Inc, MN, USA) software was used for statistical experimental design, analysis of the results and drawing of the contour plots. The model was fitted by multiple linear regression and selected by examining the 'Lack of Fit tests' in which low 'ROOT MSE' and 'PRESS' values describe the fitness of the model.

Physiology of **L. rhamnosus** *MTCC 1408 Wild Type*

The batch fermentation kinetics of the wild type strain of *L. rhamnosus* was studied at 40°C and pH 6.2 under anaerobic conditions initial glucose concentration of 100 g/l. Cells grew exponentially up to 18 hrs and the cell growth ceased after 24 hrs.

Lactic acid was produced to 50 g/l during the exponential growth of the cells but it continued up to 34 hrs to a final concentration of 80.3 g/l. After 24 hrs, the lactic acid production rate is almost constant. Lactic acid production is known to be growth associated and non-growth associated production. The production of ethanol has also shown the similar pattern. There observed a rapid increase until 24 hrs, and thereafter constant rate of production is obtained. Ethanol production also ceased with cell growth at a final concentration of 4.1 g/l. The biomass concentration was found to be 9.86 g/l. The specific growth rate was observed to be 0.31 h^{-1}.

The anaerobic glucose metabolism in *L. rhamnosus* MTCC 1408 produces lactic acid and ethanol as the fermentation product. Other products like pyruvate, acetate, acetaldehyde were found absent in the fermentation samples which is clearly observed in gas chromatograph. This has suggested us that selection of *L. rhamnosus* to be used for anaerobic L(+) lactic acid production is very suitable. The anaerobic batch fermentation profiles have also shown the little substrate diversion for ethanol (another metabolite of anaerobic fermentation). Although, major portion of this produced ethanol can be separated by physical purification/separation techniques, but for diverting the whole carbon substrate to L(+) lactic acid via anaerobic fermentation, the development of *adh*$^-$ mutant was required to produce. Moreover, the presence of even slightest amount of ethanol is hampering the polymerisation of L(+) lactic acid which was the set for the present work.

Selection of Stable Mutants

Cells were exposed to N-methyl, N-nitro, N-nitrosoguanidine (MNNG) solution of 100 mg/ml, 75 mg/ml and 50 mg/ml to develop *adh* mutants that cannot produce ethanol. When 100 mg/ml MNNG was used for 20, 30 and 40 min, 3 colonies were obtained on allyl alcohol plates. These colonies were stable upto two transfers on allyl alcohol plates. When 75 mg/ml MNNG was used for 30 min, 3 colonies were obtained on allyl alcohol containing plates.

These colonies also lost their ability to grow on allyl alcohol containing agar plates after 2 or 3 transfers on agar plates containing allyl alcohol. With 50 mg/ml MNNG, cells from broth obtained at different fermentation time (exhibiting optical density 0.63, 0.81, and 1.24), were exposed for different time intervals (20, 30 and 40 min). From this study, 36 mutants were isolated, out of which 12 mutants were able to grow on allyl alcohol plates after 6 transfers. When the growth of the 12 stable mutants were investigated for their growth in flasks, the biomass obtained after 48 hrs showed a difference. *Lactobacillus rhamnosus* NN5 that showed the highest biomass concentration obtained after 48 hrs was selected and used for further studies.

Physiology of L. Rhamnosus NN5 Mutant Strain

The batch kinetics of the mutant strain *L. rhamnosus* NN5 was studied in a bioreactor operating at similar conditions that used in batch culture with wild type strain. Minor differences in the growth profiles of the mutant strain and that of the wild type were found. The specific growth rate had decreased from 0.31 h^{-1} for the wild type strain to 0.28 h^{-1} for the mutant strain. The wild type strain produced 4.1 g/l ethanol.

No yield of ethanol was observed in fermentation by the mutant strain. The lactic acid concentration increased from 80.3 g/l for the wild type strain to 85.6 g/l for the mutant strain. However, the overall productivity for lactic acid dropped from 2.38 g/l/h for the wild type strain to 2.14 g/l/h for the mutant strain. This observation could be due to the decrease in the observed maximum specific growth rate. Inspite of a decrease observed in lactic acid productivity, the lactic acid yield was enhanced by 6.6 per cent clearly indicating the substrate amount which was diverting for the ethanol in wild type, being also utilised for L(+) lactate formation in the mutant strain.

Although, the mutant showed the capability to produce L(+) lactate as the only product in complex glucose limited medium, but the proteins present due to unutilised yeast extract were found inhibitory for L(+) lactic acid polymerisation. Therefore, further efforts were required to be made in order to minimise the protein contents in fermented medium.

Optimisation of Glucose Limited Defined Medium by Response Surface Method

In order to find the optimal concentrations of amino acids, vitamins and trace metals in the glucose limited defined media for lactic acid production, a response surface method was applied for the mutant strain *L. rhamnosus* NN5.

The intermediate metabolites of tricarboxylic acid cycle are responsible for the formation of amino acids, vitamins, purines and pyrimidines. During anaerobic fermentation pyruvate gets converted to lactic acid or ethanol and does not enter the TCA cycle. Therefore, intermediates have to be added externally to account for the same. In order to investigate the effect of supplementation of succinate, acetate and citrate, flask cultures were conducted by adding one component and excluding the other two or by adding two and excluding one of them. The addition of citrate showed the cell concentration was the highest followed by acetate and succinate showed the lowest cell concentration. Also acetate with citrate (each at 1 g/l concentration) gave the same highest optical density of culture as that of acetate, citrate, and succinate (each at 1 g/l concentration). Therefore, the concentration of acetate with citrate (each at 1 g/l concentration) was taken in the medium for subsequent experiments.

Fermentation Physiology of Mutant Strain in Optimised Glucose Limited Defined Medium

To investigate the performance of the mutant strain of *L. rhamnosus* in the optimised medium, fermentation operation was carried out using initial glucose concentration of 100 g/l. The fermentation run was performed at controlled pH of 6.2.

Cells reached their exponential growth phase after 4 hrs and reached their stationary phase by 20 hrs. Lactic acid was produced to a concentration of around 45 g/l with cell growth but continued to be produced to a final concentration of 85.6 g/l up to 45 hrs. The batch kinetics of the mutant strain in the optimised synthetic medium and the glucose yeast extract medium are comparable. For the glucose yeast extract medium, the rate of substrate consumption has been observed to be maximum during the exponential phase of growth, but wherein lactic acid was also being produced exponentially for the mutant and the wild type strain. Ethanol was also being produced by the wild type strain during the exponential growth of the cells, whereas in the optimised defined medium, a steady consumption of glucose was observed. Although, the specific cell growth rate in this medium (0.23 h^{-1}) was lower than the case with the complex medium (0.28 h^{-1}), the maximum lactic acid concentration of 85.6 g/l could be obtained. The biomass and the lactic acid yields were similar in the two media for the mutant strain. However, the volumetric productivity is found to be 8.9 per cent less in the case of optimised medium than the glucose-yeast extract medium. This was observed due to the lowering of observed maximum specific growth rate in defined glucose limited medium.

The mutant *L. rhamnosus* NN5 obtained by chemical mutagenesis (MNNG) is able to perform stably in batch fermentation without reverting back to the wild type. The observed maximum specific growth rate of the mutant strain was 0.28 h^{-1} as compared to 0.31 h^{-1} obtained for wild type strain in glucose-yeast extract medium. A small decrease in lactate productivity from 2.38 g/l/h for the wild type strain to 2.14 g/l/h for the mutant strain was observed in glucose-yeast extract medium. An optimal defined medium was obtained using response surface method. The *adh*$^-$ mutant was able to produce L(+) lactic acid as the only product with a yield of 0.88 g/g in the optimised glucose limited defined medium. L(+) lactic acid is a monomer for the biodegradable poly L-lactate polymer industry. The production of L(+) lactic acid can further be improved by various process strategies for fermentation and downstream processing (Srivastava *et al.* 1999).

Mapping Human Variations

Genetic variation research has caused problems for indigenous populations around the world.Basic human rights protocols such as informed consent are often neglected in the field. In addition, research that targets specific populations can impact those groups, although there are no standard protocols to protect group interests. Indigenous peoples have been thrust into the forefront of the new 'gold rush' for human genetic variation. Instead of land or natural resources, this time the commodity being sought from indigenous people is the unique genetic heritage flowing in their blood.

The genomics industry is really seeking genetic variations linked to diseases and other traits. The industry hopes that these variations can be translated, by obtaining patents on the variations, into profitable new products. For example, when a variation was thought to be linked to breast cancer, the variation was patented and a test to detect that variation devised and marketed.

The rush to obtain and patent such variations creates many problems for indigenous peoples. First, a high demand for unique indigenous genetic samples has resulted in researchers frequently employing less than scrupulous means to obtain them. The mandate of obtaining fully informed consent before taking samples from individuals is sometimes ignored; tribal governments frequently are not consulted prior to designing and funding studies targeting tribal members as subjects, even though the tribe will be identified in the research findings; DNA may be patented for commercial uses without the tribal or individual subject's permission or knowledge. But there are virtually no legal protections to invoke when ethical violations occur.

A particularly common practice involving indigenous genetic samples is widespread secondary use of the samples.

Once a sample is taken, it can be preserved for later uses and for a host of other studies, unrelated to the original use and without the donors' knowledge. Frequent culprits in such secondary uses are evolutionary population geneticists interested in studying relationships between different populations and migrations of populations over time. Indigenous peoples are the primary subjects for scientists conducting this type of study. However, it is not clear how such studies benefit the subjects of the research, as indigenous nations already have accounts of where they come from and who their closest relations are. In addition to violating informed consent standards, such research is also unlikely provide any benefit to the subjects of the research.

Claims by some geneticists that they can identify who is 'Native American' in genetic terms are also causing problems among Native Americans. Recently, in Vermont (USA) state legislation was proposed that would require genetic testing to determine who is Native American. In reality, there is no unique 'Native American gene'. At best, genetics could show if two people are likely to be related to one another. It certainly cannot show whether one is culturally affiliated with a Native American tribe. Even if it were possible, such a genetic approach is an affront to tribal sovereignty. Traditionally, it is the right of a tribe as a sovereign nation to determine their own members.

The level of awareness among indigenous peoples of genetic research and its impacts on our communities and environments is rising. The prominence of genetics in the recent headlines helps to increase this, although community education is necessary to overcome the press reports' gushing optimism for the promises of genetics in curing the world's woes. Although indigenous people are at the forefront of this new technology, their communities are only just beginning to realise what this means. Tribal governments are taking protective measures at the tribal and the federal policy levels, while individuals are taking their own steps, from educating community members to working for protective tribal and federal policies.

Effects of Biotechnology on Agrobiodiversity

Since the *Convention on Biological Diversity* (CBD) was signed in 1992, the value of biological diversity in general, and agrobiodiversity in particular, has gained recognition worldwide. Agrobiodiversity can be defined as the total of components, structure and functions in agro-ecosystems relevant for agricultural production. Agrobiodiversity is first of all of vital importance for the food security of future generations. This diversity can be exploited to overcome new pests and diseases, to cope with climate changes and a growing world population, to react to changing consumer demands and to make production more sustainable. Furthermore, agrobiodiversity is related to cultural values and traditions. Finally, many believe that diversity has an intrinsic value and that mankind is responsible for the survival of ecosystems and all that constitutes these systems.

Technology in general has had an impact for ages on the development of agrobiodiversity. Utensils in agriculture, food processing technology and seed storage systems, for example, have always been developed in close interaction with the diversity present in agriculture. This includes the genetic diversity of landraces and varieties within crop and animal species, the diversity of different crops and animals at the farm and the diversity within the entire agroecosystem. Some technologies, like fermentation technology, have influenced the properties that farmers consciously selected for their crops and animals. Others, like seed storage and fodder production, have resulted in conscious as well as unintended selection in landraces and animal breeds. The adaptation of crops and animals to the available technologies has determined their role in the farming system. And finally, the development of the farming systems has greatly influenced the agroecosystems in which they function, as is evident from the current discussions on the sustainability of high-external-input systems.

Biotechnology, like all preceding technologies, has influenced and will further influence agrobiodiversity. In this article an inventory will be made of the likely or potential effects of modern biotechnology on agrobiodiversity.

Common Effects of Agriculture and Biotechnology on Biodiversity

Both agriculture and biotechnology make use of biodiversity and thus show us the value of biodiversity. Likewise, both agricultural practices and biotechnology impose a threat to the existing biodiversity on which they depend.

Modern agricultural practices, stemming from the rise of a modern breeding industry and from the Green Revolution, have caused massive genetic erosion, the disappearance of many diverse populations of crops maintained by farmers and adapted to local circumstances. The application of modern biotechnology may result in a wider use of genetic diversity, whether present in wild or domesticated species, for the benefit of future food security. However, it may also potentially result in the further narrowing of the genetic base of our food crops, because of the high costs of biotechnology and, consequently the tendency to focus on few varieties or breeds only. It may even result in the introduction of novel organisms which form a risk to the (agricultural) environment.

Genetic erosion, now commonly regarded as a negative development, was caused by the rise of the modern breeding industry and marketing decisions of that industry, and did not form a natural necessity. The socio-economic context in which biotechnology is developed and applied includes competition, research structures, seed markets and intellectual property rights. It will also determine which applications and, consequently, which effects of biotechnology on the conservation and utilisation of biodiversity will dominate. For a number of biotechnologies, the picture of positive and negative potentials, depending on the socio-economic context, begins to appear.

In vitro Technology and Reproduction Technology

Tissue culture of crops offers the possibility of rapid propagation of large numbers of genetically identical plants. The technology can be used for the propagation of desirable genotypes of vegetatively propagated plants. It is applicable in particular in the case of highly heterozygous clones, of which the genotypes are lost in sexual reproduction, and for plants with decreased fertility or long generation cycles. Tissue culture is therefore, popular in horticulture and forestry. The technology can also be used to lower the virus pressure in starting materials.

These applications may increase the availability of genetically diverse starting materials. Moreover, tissue culture can be used for conservation of genetic diversity. However, if the technology is mainly used to supply large areas with genotypically identical materials, the net effect of the technology on genetic diversity will be negative. At present, the net effects are still difficult to measure. However, for some crops which have already been reproduced by conventional clonal propagation techniques for many years, the trend is clearly negative. In the case of banana, only a few varieties are dominant worldwide.

Inbreeding of properties from distant relatives using in vitro methods can result in novel combinations of genetic diversity. In practice, this option is not very popular in breeding because of the long generation times and high costs related to the introduction of genes from distant relatives. It has mainly led to rarities like a hybrid between potato and tomato and the 'shoat', a hybrid of sheep and goat.

Cryopreservation, the preservation at temperatures as low as -196°C, can be used for conservation of plant meristems, but until now by far the most important application has been the conservation of sperm of animal species. Effects on genetic diversity of this application in artificial insemination have been mainly negative: the very successful bull Sunny Boy has produced the sperm for nearly 1 million inseminations in the Netherlands, which is over 75 per cent of the total, and left a massive genetic impact on Dutch livestock.

Multiple ovulation, in vitro fertilization and embryo transfer in combination with cryopreservation may allow the rapid propagation of selected animal breeding material. This results in the rapid replacement of heterogeneous, locally adapted populations by homogeneous stocks, as has happened in the case of chicken.

Apomixis and Terminator Technology

Apomixis, the seed formation from unreduced egg cells without fertilization, occurs in nature in some plant species like cereals. The resulting offspring is genetically identical to the parent. For commercial breeding this development is of little interest since it may create a way to surpass the biological property protection offered by hybrid varieties. It offers farmers the possibility to produce seeds for the next growing season with the same properties as the parent plant, which until now cannot be realised using hybrids.

However, modern plant-breeding industry may move into crops largely ignored so far, using a new technology, called terminator technology. The American cotton seed company Delta and Pine Land and the United States Department of Agriculture (USDA) announced they had received a patent on a technique that genetically disables a plant to set seeds that germinate when planted again. This technology can potentially be used in all cultivated crops. Varieties of crops like rice, wheat, sorghum and soybean, which could not effectively be marketed using hybrids may now be commercially protected by using this new technology. This would open some of the world's largest food crops to profitable breeding and further increase the risk of genetic erosion, which has already progressed extensively in these crops. The net effect of this technology is likely to be negative.

Biological Test Kits and DNA Marker Technology

Using biotechnology, ready-to-use and simply applicable test kits have been developed for identification of viruses, fungi and other pathogens in the field. These kits can be implemented as precision instruments for a rational selection of control strategies and moments in integrated pest management.

As compared with traditional pesticide use, integrated pest management has a positive effect on the total biodiversity in agroecosystems. Similar kits applied in diagnosis of animal diseases may remove barriers between nations in the exchange of breeding material for livestock and this could also increase the availability of a wider array of breeding material. The likely net effect of biological test kits on agrobiodiversity is therefore, positive.

Many agronomically important traits are governed by a large number of genes which interact at different regulatory levels. These traits are called quantitative traits. Traditionally, manipulation of the genes involved forms the domain of breeding.

Molecular markers enable localisation of the genome sites involved in such properties and this knowledge may render breeding faster and more precise. Marker technology can be applied for the exploitation of genebank *ex situ* collections as well as for the improved use of elite breeding materials. It can be used for quantitative traits such as yield, mineral use, food quality, taste, storage capacity, resistance and adaptation to physical stresses.

Considering that most commercial crop varieties currently exhibit a narrow genetic basis, marker technology offers the opportunity for rapid corrective measures in case of outbreaks of new pests and diseases. Furthermore, the technology can be used preventively to determine the relationship between genetic materials and genetic properties and enables rational choices aiming at an increase of genetic diversity.

In animal breeding, determination of genetic relationships in animal populations is a major application. This allows a rational selection of reproduction schemes which best maintain the genetic diversity present in the population.

Finally, the technology can be used for fundamental research on the relationship between the level of genetic diversity in a crop and its versatility in reaction to changing conditions. Marker technology will surely be used by industry to improve current breeding practices, but it is uncertain to what extent applications will be realised that benefit maintaining or increasing biodiversity and versatility in agriculture.

Genetic Modification

Genetic modification has been applied in both plant and animal breeding, although commercial applications have only been realised in the former domain, e.g. insect resistance and herbicide resistance in maize, cotton and soya bean, and prolonged shelf life of tomato. In view of the high costs of genetic modification, the technology will only be applied in a few economically important crop varieties of major crops grown in the industrialised world, and possibly in animal breeds. If the resulting genetically modified organisms form an agricultural success this may lead to replacement of several conventional varieties and thus to a decrease in genetic diversity in that crop or animal. Furthermore, patent protection will limit the use of such varieties in further breeding by others and thus prevent the spread of the newly introduced traits. Although patent protection is not a necessary consequence of biotechnology, its result is that the net effect of genetic modification on biodiversity is likely to be negative. This is contrary to popular statements about the addition of novel genes to available crop genotypes.

Genetic modification of microorganisms with an agricultural value such as genetically modified Bacillus thuringiensis may improve bioinsecticidal sprays or enlarge their host range. Modification in relation with nitrogen fixation, which can reduce the use of artificial nitrogen fertilizers, offers only very limited opportunities in the short run because of the intricate biochemical pathways involved. Positive effects on biodiversity stemming from the reduction of the use of artificial fertilizers is therefore, unlikely.

Genetically modified veterinary vaccines offer an opportunity to increase the health of all sorts of livestock, be it indigenous or exogenous, common or rare and net effects are hard to predict. However, increased use of more effective vaccines might fit the tendency to control the environment rather than to exploit genetic diversity in the search for better production.

A Positive Role of Biotechnology in Maintaining and Enlarging Agrobiodiversity

Is the picture of the effects of biotechnology on agrobiodiversity indeed so gloomy? Much will depend on the ways biotechnology is exploited, and positive applications are certainly feasible. To summarise, what is the potential of biotechnologies to stop and reverse the decrease in diversity in agriculture?

Maintenance of Genetic Diversity: Marker technology can be used to study relationships between genetic material, in starting materials for breeding and in small animal populations, and to monitor the genetic developments in *in situ* and on-farm conservation projects.

Cryopreservation increases the options for conserving starting material, in particular for animals and for those crops which cannot be maintained using normal practices.

Utilisation of Genetic Diversity: Markers can also be used to increase the efficiency of breeding strategies and to focus on desirable traits. Since markers allow precise identification of the genetic make-up of varieties, they allow relaxation of legislation on uniformity of crop varieties and may support reintroduction of genetically more heterogeneous and versatile varieties.

Finally, markers can be used to trace properties which are important in ecological farming or other forms of sustainable production which are currently under-utilised. Cryopreservation in combination with artificial insemination may help to support the maintenance of rare breeds or rare properties.

Reduction of Pesticide Use: Integrated pest management using test kits and resistance breeding using marker technology or (in cases of durable resistance) genetic modification may reduce the need for pesticides with an overall positive effect on biodiversity.

Overriding Socio-economic Context

In many countries, government policies and efforts of private breeding companies have stimulated the shift from food production for subsistence or for the local community in complex farming systems to production of fewer crops and breeds for the national or global market. Often this has implied a shift from many local, well adapted landraces to few modern high-external-input varieties.

These policies and market efforts have resulted in a concomitant loss of diversity. These effects occurred in most developed countries but have now also become highly significant in most developing countries. Opening markets and globalising economies, and the efforts of the international and national breeding institutes on major food crops are responsible for these effects. Higher food production based on high-external-input varieties and loss of traditional genetic diversity which we may need in future are two sides of the same coin.

Modern biotechnologies will strengthen these effects. Tissue culture techniques and artificial insemination require relatively low investment costs and limited expertise and are now well within the reach of most developing countries. Marker technology and genetic modification clearly require higher inputs, better facilities and more expertise, but are applied by several developing countries. And certainly, institutes which focus on agriculture in developing countries, e.g. the *Centro Internacional de Mejoramiento de Maiz y Trigo* (CIMMYT), the *International Rice Research Institute* (IRRI), the *Centro Internacional de la Papa* (CIP), the *Centro Internacional de Agricultura Tropical* (CIAT) and the *International Livestock Research Institute* (ILRI), have acquired these latter techniques and are involved in applications which will render novel breeding products for agricultural production. Only the *International Plant Genetic Resources Institute* (IPGRI) and the *Centres' System Wide Initiative on Plant Genetic Resources* of the *Consultative Group on International Agricultural Research* (CGIAR) are unambiguously dedicated to the rescue of agrobiodiversity that has remained. Future breeding strategies within the CGIAR Centres will probably have a major impact on conservation and utilisation of genetic diversity in farmers' fields in developing countries.

In the CBD, loss of biodiversity in general and agrobiodiversity in particular has been recognised as undesirable. Improvement of production under low-external-input conditions, long-term sustainability, and future versatility of agricultural production are all dependent on a wide and rational utilisation of agrobiodiversity. In modern agriculture, agrobiodiversity has received too little attention.

Genetic diversity was regarded as being functional in breeding programmes, but not in the field where conditions are controlled and where uniformity is required for high yields. Development and application of technology in agriculture took place from this perspective, and largely ignored negative effects on agrobiodiversity. Conservation of biodiversity for the food security of future generations and for a more sustainable agricultural production requires a change in this attitude. It also requires a change in socio-economic conditions in agriculture which have caused the neglect for maintenance of biodiversity.

This change should be based on the acknowledgement of the value of *in situ* conservation of genetic diversity and the need to consciously develop and use technology in this direction. This change should form a challenge for both governments and industry when opening new markets. Only then will the net effect of biotechnology on biodiversity be positive.

India's Biotechnology Plans

On paper, India's new biotechnology plans for the next millennium seem feasible. However, given the broad coverage that these plans envisage, along with the lack of public-private linkages, questions can be raised about successful implementation.

The plans have been designed by a group of experts, including both scientists and industry representatives, who were appointed by the government. They were published in the *Report of the Working Group for the Ninth Five Year Plan* (1997 to 2002), and contain both short-term and long-term, i.e. longer than five years, research and production goals. With a proposed expenditure in excess of US$ 255 million over the five year period, the experts have represented the most ambitious plans ever drawn up in India for this technology. Although, compared to previous years, the absolute budget for biotechnology has increased, in relative figures biotechnology research faces a declining share in the overall scientific research that is carried out in the public domain. While the government's budget for scientific research has increased by 32 per cent, the share of biotechnology has declined from 11 per cent in 1987-88 to 9 per cent in the current financial year.

The basic policy framework for biotechnology has two main dimensions. The *first* takes into account the priorities of the main sectors in which biotechnology can make a difference. The *second* includes the discussion on some of the institutional mechanisms that are crucial to the developments in this sector, for instance the issue of *intellectual property rights* (IPRs). By accepting membership of the *World Trade Organisation* (WTO), India has committed itself to extending intellectual property protection on living matter, for instance microorganisms and plant varieties. Another area of institutionalisation is to find regulatory mechanisms for biosafety.

Main Objectives

The priority areas have been identified by the experts based on two objectives: *Firstly*, it has been proposed that the programmes in the various sectors must lead to the generation of additional employment. *Secondly*, the needs of rural women and people from the economically weaker sections of the society have to be kept in sight. Therefore, a number of existing demonstration and training centres are sought which can be strengthened for transferring simple and low cost biotechnologies for economic improvement of these communities. For example, a biotechnology park is being planned in the Southern Indian state of Tamil Nadu. This has been conceived as a commercial venture with the participation of the *Department of Biotechnology* (DBT) and the *Tamil Nadu Industrial Development Corporation*. Among the technologies selected for this venture are tissue culture, biological pest control agents and organic compost.

In the report it has also been proposed to establish an advanced centre for biological pest control research. This centre would have regional centres to impart regular training to young scientists, farmers and unemployed youth on, for example, the safe use of biotechnology. In addition, the centre could develop technology packages and would act as a repository for the collection, maintenance and supply of biological pest control agents.

Agriculture

The priorities set in the report with respect to the agricultural sector reflect a continuity with past activities. However, there have been some changes. The proposal for the budget for crop biotechnology shows a significant increase: between 1992 and 1997 about US$ 5 million was spent, while a proposed budget of more than US$ 17 million is to be spent in the ensuing five years.

The report prioritises the use of biotechnology for both increasing yields and for the diversification of crops presently under cultivation. Also, higher crop quality, such as higher nutritional value, is targeted. For instance, DBT has a programme for improving the nutritional value of pulses. In addition, genetic enhancement centres should be established to apply DNA marker technology for crop improvement and the identification of pathogens and pests. The crops identified for this purpose are cereals, pulses, and oil seeds. Besides food crops, attention is also to be given to the improvement and mass propagation of plants for traditional medicinal use, horticultural plants and endangered species.

In the areas of sericulture, which rears silkworms for silk production, and aquaculture, it is proposed that the ongoing programmes for developing diagnostic kits for early detection and prevention of infectious diseases are strengthened. This has become necessary since, due to the growth and intensification of aquaculture, there is increased incidence of diseases in the cultivated organisms.

Medical Biotechnology

In this area, the allocation proposed in the Ninth Five Year Plan is US$ 18 million, as against US$ 7 million spent in the period 1992-97. This illustrates the ambitions of the plans that have been developed.

The proposed prioritisation of the programmes is based on several criteria, including the degree of disease-burden, mortality rates, the cost effectiveness of treatment, the emergence and re-emergence of diseases and the potential use of technology at the community level. It is planned to develop cost-effective and quicker diagnostics of tuberculosis, HIV, leishmaniasis, respiratory tract infections, cancer, and rheumatoid heart diseases.

Vaccine development has been one of the principal areas of biotechnological research in medicine. In the past, government initiatives towards vaccine development focused mainly on tuberculosis, HIV, cholera, typhoid and rabies. Now, malaria has been added as a disease that has made a come-back after having nearly been eradicated over the past few decades. The specific problems associated with malaria in India are a significant increase in the recorded mortality rates and the ineffectiveness of conventional drugs due to increased drug resistance.

An added dimension to the proposed research is the development of transgenic animals for biomedical purposes. These transgenic animals would serve as biofactories for biomedical products or as experimental modules for research on human genetic disorders and AIDS.

Bioinformatics

Among the proposed initiatives is the development of an effective bioinformatics infrastructure through the establishment of a *National Bioinformatics and Biocomputing Institute* (NBBI) to coordinate the activities of the large number of public bioinformatics centres that have been operating in the country. A *Biotechnology Information System* (BTIS) has been established by the major university departments and public sector R&D institutions to fulfil the growing need for information related to biotechnology. It works as a distribution database and network organisation to provide integrated information on all aspects of biotechnology including:

- Genetic resources data, such as germplasm databanks, DNA and protein sequences.
- Bibliographic references.
- Management information, for instance on R&D projects and resource directories.

Bioinformatics is seen as a useful mechanism considering the vast volumes of genetic and biological data emanating from the research laboratories. Since major transnational corporations will increasingly use bioinformatics to make their operations more effective, the NBBI might become an important means of attracting foreign investment, especially from transnational companies.

Limitations

Lack of Links: Links between publicly funded research institutions on the one hand and local industry and civil society on the other, are weak. For example, the government-sponsored biotechnology R&D institutes have developed 17 technologies for different private companies, but none has been found commercially viable by these companies. Among these technologies were detection kits for several diseases such as hepatitis B and typhoid fever, tissue culture for bamboo, and an animal birth control injection.

Lack of Focus: Another feature of the proposed policy is that every conceivable sector is included in the list of priority areas, and that the policy, therefore, lacks focus. Attention is drawn, for instance, to biodiversity conservation, environmental protection, agricultural biotechnology, medical biotechnology and bioinformatics. It is worth noting that this overload of proposed priorities contradicts the fact that the biotechnology sector faces severe resource constraints, even for the continuation of various existing programmes.

Isolation: Like earlier plans, this report neglects several relevant areas. *Mira Shiva* from the *non-governmental organisation* (NGO) *Voluntary Health Association of India* feels that the biotechnology research priorities are still developed in isolation, irrespective of the national requirements. The priorities do not address the lack of access to health care and the development of technologies that meet the requirements of the country's growing number of poor. Important food crops such as sorghum, millet and cassava do not appear on the list of priorities.

In the area of medical biotechnology, a similar hiatus between the orientation of research priorities and perceived needs can be found. The focus of research in medical biotechnology has remained fairly diffuse. Even basic requirements, such as for the polio vaccine, which is part of India's general immunisation programme, are mostly met via imports.

Procedural Problems: Long-winded procedural problems contribute to difficulties concerning the implementation of the programme. The intention is that after the priority areas of research are identified and published, scientists will be invited to send in proposals. A committee inside the institution from which the proposal comes will screen the projects to assess their relevance to the identified priority areas. The eligible projects will then be reviewed by national experts and finally evaluated by a DBT task force, consisting of experts from government departments, scientists from universities and autonomous, publicly funded research institutes. Currently, there are 16 biotechnology task forces. The screened projects, eventually cleared by the task force, will finally be reviewed by the finance section of the DBT.

This process has two underlying dangers. *Firstly,* the task forces that scrutinise the project proposals were set up in 1987, and their relevance in relation to recent developments in biotechnology has never been questioned since. For example, among the many task forces established in 1987, one focused on tissue culture propagation techniques.

Since then, several centres have become involved in this area. Therefore, tissue culture no longer needs the public support it once did, and this money could now be used for the solution of urgent problems and the development of more advanced techniques. *Secondly,* as the decision rests with the financial experts, it is likely that they will strive for an optimal allocation of resources by ensuring that each task force receives some assistance. In this process, projects with larger national goals sometimes lose, since not all task forces deal with issues important in terms of national requirements.

In summary, given the large number of activities the DBT proposes to undertake during the next five years, it difficult to see how it would be able to realise these objectives. It remains doubtful whether these activities will help to generate employment and to meet the needs of rural women and people from the economically weaker sections of society.

Proposed Changes

Several suggestions have been made to ensure that research in the area of biotechnology complements the ongoing research in agriculture and medicine. *Suman Sahai* of the Indian NGO the *Gene Campaign,* feels that there is a need to develop a comprehensive *National Bioresource and Biotechnology Policy* (NBBP). Such a biotechnology policy should clearly articulate the biotechnology goals of the country and identify the problems associated with these goals which cannot be solved by employing traditional technologies. This approach, in Sahai's view, would compel the domestic R&D agencies to develop a more restricted and home-grown research agenda.

Such an NBBP would also been seen as a framework for ensuring a larger involvement by the private sector in the development of biotechnology on a national level. It has been proposed that the government provides assistance to the private sector for obtaining funds for product development. One of the mechanisms for the development of indigenous R&D that has been discussed in India and elsewhere, for instance in Singapore, is the provision of venture capital funds.

Pushpa Bhargava, the former director of the *Centre for Cellular and Molecular Biology*, one of the leading publicly funded biotechnology laboratories in India, has suggested that the biotechnology research institutions be asked if their research activities can help in establishing a competitive Indian biotechnology industry. This, Bhargava feels, can help tackle the critical problem of the absence of any accountability, whether scientific, financial or social, in biotechnology research.

The lack of accountability could be addressed by the involvement of NGOs in the formulation of the national biotechnology policy. At present, however, this is not the case. Several factors have contributed to this situation.

Firstly and most importantly, there are some deficiencies that the NGO sector faces at the present juncture. They are still small in number and most organisations' capacity to intervene is restricted to their own geographical environment. In fact, the effective functioning of the NGOs could have far larger benefits. The beneficial role that these organisations can play was recently confirmed by several cases in which government research institutions violated biosafety regulations. The NGOs indicated the illegal use of the *Bacillus thuringiensis* (Bt) gene in field trials.

Secondly, the DBT has not made any concrete effort to provide a forum for the NGOs. No effort is made to obtain feedback from NGOs on biotechnology policy in general and on the R&D priorities in particular. NGOs have a rich experience arising from the work with people at the local level and their expertise should help to identify areas in which biotechnological intervention could be rewarding.

Tissue Culture for Coffee: The Case of Uganda

Many African countries have started biotechnology research to increase the competitiveness of their major cash crops. Some of these countries are dependent on one or two crops for their foreign exchange earnings, and the application of biotechnology to enhance their competitiveness might become essential to their participation in the world market. The case of coffee biotechnology research in Uganda, where researchers are trying to enhance the country's competitiveness in the world coffee market by increasing the stock for replanting, illustrates this point.

The Importance of Uganda's Coffee

Uganda has gone through two decades of political turmoil, economic decline and agricultural transformation. Since the coming to power of President Museveni in 1986, the country has embarked on a major economic recovery programme, which has resulted in a significant increase in economic performance reflected in an average annual growth rate of nearly five per cent.

Uganda is Africa's second largest coffee producer after Côte d'Ivoire. Coffee is its main export crop, in 1990 accounting for over 79 per cent of foreign exchange earnings. In terms of total domestic revenue, it contributed about 12 per cent in the form of export tax. Foreign exchange earnings decreased in the early 1990s due to the collapse of international coffee prices. The coffee export earnings of 1992 reached their lowest level of US$ 98 million, compared to the average annual earnings over the period 1982-1989 of US$ 300 million. The share of coffee exports in total exports dropped to about 55 per cent in 1993, while export of non-traditional products increased about 13 per cent between 1992 and 1993.

Coffee production is spread throughout all of Uganda. The robusta variety accounts for up to 92 per cent of total output. The production of coffee uses 250,000 hectares out of the country's 23.6 million hectares. The entire production is carried-out by approximately 1.2 million small-scale farmers, nearly 7.5 per cent of the population and 50 per cent of the small-holder population.

Although industrial output has been growing at a rapid rate of over 10 per cent annually in the last three years, Uganda still has one of the highest ratios of dependence on one cash crop in Africa. Its economy is, therefore, vulnerable to changes in international coffee prices, or to technological innovations that might enable the growing of coffee in countries which are not among the traditional producers.

Increasing Coffee Production

The replacement of old coffee trees, yield and quality improvements, and enhanced disease resistance of coffee have been a major policy area in Uganda. In 1990, the *Makerere University* in Uganda started a US$ 10 million programme to introduce clonal coffee trees in the country. The programme's finances were provided by the Uganda government and the *European Union* (EU). The aim is to replant all the farms by the end of 1995, providing sufficient financial resources.

About two years ago, it was decided to concentrate all research efforts and funds at a government research institute, the *Kawanda Agricultural Research Institute* (KARI), where research on tissue culture in coffee had also been conducted. Currently, the project is still funded by the EU and the government, while training and technical support is provided by the *Centre for International Cooperation in Agricultural Research for Development* (CIRAD, France). At present, both KARI and the *University* are conducting research in coffee tissue culture.

The yields of robusta in Uganda are around 600 kg of clean coffee per hectare, which is estimated to be 75-80 per cent of the potential yield levels using low-input techniques. The minimum yields for arabica are 450 kg per hectare, some 60-75 per cent of the expected output under small-holder conditions. Six clones of robusta coffee capable of giving yields of 2 to 3.5 tons per hectare of clean coffee with fertilizer application have become available since the mid-1970s.

For example, clonal robusta varieties were developed in Uganda in the 1970s, but were not introduced because of the political turmoil of that time. Without fertilizer application, this robusta variety yields up to 1,100 kg of coffee per hectare, and is resistant to the coffee berry disease. Additionally, all six clones have large berries, ripe uniformly, and have a good quality. However, the farmgate price is still as high as that of the other varieties and only a small quantity of these six clones has been available from the coffee nurseries. The main reason for this is the absence of an available optimum propagation technique.

Coffee Tissue Culture

Conventionally, coffee is propagated from seed or by vegetative cuttings. Both these methods of propagation have disadvantages. Seed propagation is associated with inherent uncontrolled genetic variation in heterozygous cultivars, slow rates of multiplication of seed, and short span of seed viability. Propagation of coffee by vegetative cuttings guarantees uniformity. Since there is more uniformity within plants of the same clone than within plants of different clones, farmers are advised to plant mixed clones in their fields and not to use a single clone (controlled variation).

Cuttings generate relatively low multiplications rates as they can only be obtained from upright (orthotropic) branches. Also, rooting on vegetative cuttings can be very difficult without the application of rooting hormones. At present, approximately 8,000 plants can be produced per annum using this conventional propagation method. This is insufficient to meet the demand.

Multiplication by tissue culture techniques could provide a viable alternative to these traditional methods of coffee propagation. Tissue culture methods permit the production of relatively uniform plants on a massive scale in a shorter period, and with a narrower genetic base than is possible under the conventional methods. Again, the danger of this narrow genetic base is addressed by mixing clones. Using tissue culture for the propagation of coffee plants, three different methods can be used: micro-cuttings, direct somatic embryogenesis, or plant regeneration from callus.

Micro-cuttings or nodal culture comprise a tissue culture approach which entails culturing nodal stem segments carrying dormant auxiliary buds, and stimulating them to develop. Since this method involves the exploitation of buds already present on the parent stock plant, it provides means of clonal multiplication. Each single segment can produce 7-9 micro-cuttings every eighty days.

However, coffee nodule culture has high primary contamination rates ranging between the 60 and 100 per cent, depending on the age of the nodal segment cultured. This contamination is often due to fungi which emanates from the mother plant in the open fields.

Furthermore, coffee contains high levels of phenols, making phenolic oxidation a serious problem in primary root cultures. Phenolic oxidation is the browning of plant tissue culture and the adjacent culture medium due to the light-mediated oxidation of polyphenols to quinines which are highly toxic to the cultured tissue. It has also been noted that the rate of phenolic oxidation is dependent on the source of nodal segment and species, robusta coffee being more prone to phenolic problems than arabica coffee.

The problem of high contamination rates can be reduced by pretreating the mother plants with fungicides prior to the collection of the primary micro-cuttings. Also, micro-cuttings of the three nodes from the apex are normally used, since the rate of contamination tends to increase with the age of the nodal segment. In this connection, young plants grown in greenhouses are preferred as sources of micro-cuttings.

A promising solution to the problem of phenolic oxidation is pre-soaking of micro-cuttings in anti-oxidants such as citric acid and ascorbic acid prior to culture initiation. Anti-oxidants are also incorporated in the culture medium, while activated charcoal is incorporated in the medium for absorption effects.

At the first sign of phenolic oxidation, the cultured nodule segments are often transferred to a fresh medium. In spite of all these precautions, great difficulties have been experienced with robusta coffee in laboratories in producing plants through micro-cutting. Therefore, before being properly exploited as a means of propagating Robusta coffee, it needs further investigation.

Direct somatic embryogenesis in coffee is normally induced by culturing leaf segments in a medium until the development of well-formed somatic embryos takes place. Direct somatic embryogenesis has several advantages over propagation by micro-cuttings. It does not have the problems of phenolic accumulation in the medium, and the contamination rates are generally very low. Furthermore, the genetic variability is relatively low at 5-10 per cent and the method permits the development of plantlets in larger numbers than is possible with the micro-cuttings method. For example, one leaf segment can produce hundreds of plantlets in a single culture cycle, whereas the micro-cutting method requires several sub-culturing cycles in order to get a comparable number of plantlets.

In cloning local robusta coffee, the *Department of Crop Science* at the *Makerere University* has now successfully obtained plantlets with this method. After 10 weeks of culturing, torpedo-shaped embryoids were transferred to an appropriate regeneration medium where they developed into plantlets in four weeks. The plantlets were acclimatised, but the survival rate has remained low. More investigations on acclimatisation techniques are needed before this method is to be commercially exploitable.

Propagation through *callus regeneration* is normally done

by inducing it on leaf segments. After four weeks of culturing leaf segments on the appropriate medium, callus starts to form at the edges. The developed callus is then transferred to a germination medium and develops numerous embryoids in 5-6 weeks. The embryoids start differentiating into a first pair of leaves in another two weeks of growth. They are transferred to a germination medium where they develop into plantlets. The plantlets are acclimatised in the greenhouse, but the survival rate is again very low.

This method can generate more embryoids than the direct somatic embryogenesis method, but it has a high genetic variability rate (somaclonal variation). It also takes relatively longer to develop plantlets than the micro-cuttings or direct embryogenesis methods. However, this method can be a very useful tool in developing new plant varieties by screening and testing somaclonal variants for characteristics of agronomic potential.

Although the research is in its early stages and it will take several years before the techniques are applied, it might contribute to an improvement of Uganda's position on the international coffee market. Other coffee-producing countries, such as Kenya and Colombia, are also conducting research on coffee, but they are generally putting more emphasis on arabica than on robusta. Such technological responses, however, will not safeguard the long-term stability of the economy. There are too many other competitors entering the market and the prices are generally highly fluctuating. Economic stability will, therefore, depend on the economic diversification currently being planned.

Significant Role of Biotechnology

The reduction of hunger and poverty is a prerequisite for sustainable development.

Malnutrition on a global scale is not only a problem of distribution (purchasing power and availability) but more and more becoming a problem of production and quantity. In order to ensure the food for a growing world population and to meet the increasing demand for high quality food, measures for the increase of food production are needed – apart from economical and social interventions. In the future it will hardly be possible to increase the area under cultivation, therefore a focus must necessarily be on intensifying crop production and technological innovations.

Due to technological advances during the "Green Revolution" it has been possible to increase agricultural production considerably over the past decades. However, these developments – promoted by the massive increase in world population – have caused a heavy strain on and degradation of natural resources, leading to the question to what extent common technologies alone can achieve the demanded increase for agronomic production. An expansion of current technological instruments seems necessary in order to allow for an increase in production and to use natural resources more efficiently and preserving at the same time. In this scenario bio- and genetechnical practices are increasingly discussed in complementing available technologies.

The broad potential applications of genetic engineering and promising approaches of its realisation indicate that this technology can and will make a significant contribution to global food security, and therefore to a sustainable development as defined by the UNCED.

Although potentials of genetic engineering have been recognised its technological development and use in tropical and subtropical plant production have played a minor role perhaps with the exception of Newly Industrialising Countries like China, Argentina and Mexico. In 1997, transgenic plants have been grown worldwide on 12 million ha and after all 25 per cent of these in developing countries (James, 1997a). More recent data state an area under cultivation of 27.8 million ha for 1998 excluding China (James, 1998a).

Whether the potentials of genetic engineering can be realised for developing countries in the future is not at all a question of technology. Institutional frame conditions which support the technological system play a key role. This includes adequate mechanisms of technology transfer as well as legal regulations in the area of intellectual property and biosafety. Within the German public the discussion on biotechnology and genetic engineering is dominated by its costs, expected profit, potential usufructuaries and especially by its risks.

A major part of the population is still sceptical or in opposition to these new technologies in agriculture. Nevertheless advances in technical development have been made and produced facts. Especially in the context of developing countries, practical aspects of realisation should, therefore, be in the foreground of discussion apart from basic potential benefits and risks of biotechnology and genetic engineering.

In the evaluation of biotechnology and genetic engineering the focus must be directed to its contribution to a sustainable development. Furthermore, there is a need to distinguish which tasks in respect to technology development have to be met by the public and which by the private sector. Biotechnology research is currently dominated by the private sector. It needs to be determined to what extent there are commercial incentives for companies to direct their R&D efforts and production activities towards the problems of developing countries. Solely solving this question creates a frame for efficient usage of the restricted public funding to the benefit of the target group of producers and consumers in tropical and subtropical regions.

Potential applications of biotechnology and genetic engineering Biotechnology includes all techniques that aim to regulate and use biological processes for the production of substances and services. Areas of use can be found wherever one is principally concerned with examination, change or optimisation of living-processes. Genetic engineering in its narrow sense includes all principals of production and use involving molecular biology and molecular genetics in which the genotype of respective organisms, their DNA, is artificially isolated, amplified, characterised and/or modified.

Potential applications of biotechnology and genetic engineering, apart from agriculture and nutrition, are also found in environmental technology as well as in pharmaceutical and medical areas. Furthermore, biotechnology and genetic engineering are applied in industrial processing, developing new materials as well as obtaining raw materials. The following explanations concentrate on agriculture nutrition and give a short overview on other applications.

Agriculture and Nutrition

One aim of the employment of bio- and genetechnical procedures in agriculture is the increase of efficiency by means of improving performance and/or a reduction of production input as well as a diversification of production branches. In addition, the efficiency of methods in breeding, selection and propagation using biotechnology and genetic engineering can be increased and genetic variability in a population broadened.

The application of biotechnology and genetic engineering can lead to an increase in food production from plants as well as animals. This increase in production plus a stabilisation of yield is mainly due to improvements in breeding and more efficient methods in the propagation of crop species and domestic animals. Besides, there is an increasing interest in the use of genetic engineering in food processing.

Plant Production

Stabilise and Increase Current Yield Ptential: Yield stability can be achieved by breeding for resistance and tolerance against biotic and abiotic stresses. Furthermore, yield increase, especially in vegetatively propagated crops, can be achieved by the aid of biotechnology, provided that cell culture and diagnostic tools realise the improved potential of propagating high-yielding, pathogen-free plant material right in time.

Sustainable Improvement of Production: Increases in productivity are possible using biotechnology and genetic engineering in breeding, to an extent which are not achievable or only at over-proportional costs using conventional breeding and selection. Marker-assisted breeding, which provides for a continuous check of the recombination of genes-of-interest during breeding programmes is highly promising (Tanksley and McCouch, 1997). Next to the direct increase in yield, genetic engineering including marker-assisted breeding, can add to an efficient application of available resources, e.g. optimised nutrient-uptake and use.

Improvement of Product Quality: The quality of plant products can be increased by optimised composition and/or supplementing nutrients (e.g. increase levels of essential amino acids and vitamins). They can be adapted to the changing demands of consumers and industrial processors. An effective measure of intervention is expected from modern biotechnology assisted breeding programmes in this area, since transformation with specific genes/traits can be done without changing the genetic background of a crop significantly. This appears to be a possibility to circumvent the often experienced negative correlation of parameters for yield and quality (TAB, 1998). In addition, shelf-life and storage can be improved through biotechnology.

Increased Quality in Regenerative Raw Materials: By using genetic engineering for interfering with the many and diverse metabolic pathways that have evolved in plant species during their evolution, agricultural crops can be used as bioreactors. In this way, plant storage compounds can be qualitatively and quantitatively modified and optimised for human use. Secondary plant metabolites as well as other desired substances can also be synthesised and accumulated so that products of agricultural crops (oil, fibre, energy, medical compounds, etc.) can again compete with non-renewable resources like coal, gas and petroleum (Brandt, 1995).

Animal Production

Biotechnology procedures increasingly dominate breeding and veterinary practices in animal keeping and open up new areas of production. Thereby biotechnology reproduction techniques are most important in conventional animal keeping.

Techniques such as artificial insemination, regulating the menstrual cycle and influencing maturity, harvest and transfer of embryos, *in vitro* fertilisation as well as sex determination and diagnosis are already broadly applied techniques in agricultural practice (Henze *et al.*, 1995).

Recombinant DNA technologies are as yet only important for research. It can be assumed that optimised procedures in the areas of marker-assisted genome mapping, analysis of gene regulation, genetic diagnostics and recombination, will gain momentum in future breeding and production processes. Hereby, gene functions concerning the whole organism or showing only local response, can be discriminated. In both fields, there are as yet only few ready-to-use applications (Fischer, 1998).

The application of recombinant growth hormones (being an example for a procedure influencing the whole organism) has documented its limitations. Although the growth potential of transgenic animals has been increased, negative effects such as increased susceptibility to diseases have been observed. Genetic engineering procedures for the targeted synthesis of desired substances in animals gain importance in the production of pharmaceuticals.

Production of substances like blood coagulating factors or interferone by transgenic animals has opened new areas of production branches in animal husbandry—as for example "Gene Farming" (Müller and Brem, 1998).

One should consider that the commercialisation of transgenic animals can only be expected in medium-term (Fischer, 1998). Prioritised research and breeding objectives depend on respective productive livestock species. In cattle, milk proteins are in the foreground whereas in pigs it is an optimised utilisation of fodder and reduction of fat content. Wool quality and virus resistance are breeding aims in sheep. In poultry and fish the improvement of growth potentials are of major interest.

Food Technology

So far breeding programmes have mainly been adapted to the demands of agricultural producers (increase and stabilisation of yield, etc.). With the aid of biotechnology and genetic engineering, also the expectations of consumers and the processing industries can be considered.

In food, the amount and composition of nutrients are criteria for product quality. With a view on medical and nutritional value, animal and plant metabolites and so-called "nutraceuticals" (describing nutrients of pharmaceutical value) are of prime interest.

For the downstream-processing industries most relevant changes include those, which facilitate processing or enable the use of new procedures (e.g. modified composition of fatty acids and carbohydrates in oilseeds and potato respectively).

Furthermore gene-technology has a direct influence on food processing by means of recombinant additives and auxiliary materials. Microbiological analysis of food stuffs using recombinant DNA techniques and genetically modified microorganisms for production of fermented foods (e.g. yeast) or for the production of additives and auxiliary materials for a better processing of food stuffs (e.g. enzymes) are the three basic applications of genetic engineering in the area of food production and processing (BLL, 1995).

Other Applications

Three basic areas can be identified in which the employment of modern biotechnology and genetic engineering methods accomplish a major contribution in improving the current environmental situation:

Bioenergy: By reducing the use of fossil fuels and increasing the rate of photosynthesis the rising of CO_2-levels in the atmosphere can be slowed down. The direct or indirect use of plant biomass as an energy source can contribute significantly. Hereby genetic engineering can also be applied (Leisinger, 1990).

Plant Protection: Genetic engineering allows the production of plants that are tolerant to pathogens, resistant to diseases and efficient in nutrient use, resulting in a reduced application of persistent and toxic chemical plant protective agents and, therefore, help saving our environment (BML, 1997).

Degradation of Toxic Agents: Genetically modified microorganisms enlarge the capability of degrading persistent toxic substances in the environment as well as producing readily degradable waste materials (e.g. packaging material) (BMZ, 1994; OECD, 1997).

Furthermore, recombinant DNA-technologies – such as the increased efficiency in genome analysis – will facilitate information recognition and processing of genetic resources that are stored in *ex situ* facilities, and hence increase the economic value of this conserved material enormously (Virchow, 1997a). These advances are also likely to be of significant value for the employment of genetic resources in environmental aspects. In general it holds true that biotechnology and genetic engineering contribute significantly to a better environmental protection by means of efficient use of resources. One can imagine for example that by increasing drought tolerance in crop plants the global use of water could be reduced markedly. The technologies also help to relieve the environment, among others through decreased application of chemical plant protective agents and a better waste removal as well as increased recycling of materials. In addition the development of new catalysts based on enzymes can lead to the employment of "softer" chemical processes (Linder, 1996).

Medicine

Genetic engineering is important in several medical branches. Firstly a better understanding of cause and development of diseases can be achieved through basic research using genetic engineering. The exact understanding is a prerequisite for new therapies that do not necessarily have to be based on genetic engineering. The currently more heavily applied area is the production of recombinant human proteins as compounds for therapeutic purposes as well as the research for the production of new and more effective vaccines. In this respect approaches for developing countries that aim to fight widespread tropical diseases like malaria, bilharziosis, etc. are highly relevant. With the aid of genetically optimised microorganisms also the production of low molecular weight substances (e.g. antibiotics) can be made more efficient. Furthermore genetic engineering is applied in improved diagnosis of inherited diseases ("gene/genetic diagnosis") as well as curing such diseases ("gene therapy") in the medical branch.

Biotechnology below Genetic Engineering

Traditional processes of biotechnology like biopesticides and biofertilisers have been intensively applied for several decades in many poor regions of the world. The nitrogen fixing blue-green algae *Azolla* for example is used as fertiliser on at least two per cent of the worldwide area under rice cultivation (TAB, 1995). Often the biological preparations are products that have been developed in developing countries themselves – partly based on indigenous knowledge (Bunders *et al.*, 1996). Also tissue culture techniques are widely applied in developing countries (BMZ, 1994; Ebert *et al.*, 1994; Wolpers, 1994; Brenner, 1996). Almost all of these cases are concerned with comparably simple micro-propagation techniques for the commercial production of pathogen-free planting material of tuber plants, fruit, vegetables and ornamental plant species as well as tree cultures with long generation cycles. Through the simple handling and labour-intensity, developing countries have comparative advantages over industrialised countries with higher labour costs.

Genetic Engineering

In 1999 genetically modified crops have been grown on almost 40 million ha worldwide.

This area has increased by 12.1 million ha since the previous year. Genetically modified soybean and maize are so far distributed the furthest. They account for 82 per cent of the area. With respect to modified characteristics herbicide tolerances are dominating (71 per cent) followed by insect resistance (22 per cent) and combined (stacked) insect and herbicide tolerance (7 per cent). Virus resistance and modified production characteristics are of less importance with respect to cultivation area (<1 per cent). However, its importance can be expected to increase in the future when transgenic plants can be used as bioreactors for the production of different substances.

Even though in 1999 transgenic crops in developing countries amounted to about 18 per cent of the global transgenic area, this number overestimates the technical capacities of the partner countries in the South. There are only very few developing countries that grow transgenic crops in considerable amounts. Most other countries of the South are far from using respective products in agriculture let alone developing them.

Even within the group of industrialised countries only two are applying genetic engineering in agriculture on a commercial scale worth mentioning. However, further distribution of such technology is hindered by a lack of public acceptance rather than a lack of technical capacity. The proportion of developing countries in worldwide transgenic area has decreased from 43 per cent in 1996 to 18 per cent in 1999. According to the current acceleration in technical development within industrial countries the portion from developing countries will decrease more and more in the future.

Until 1997 more than 60 transgenic plant species and varieties have been produced in different countries of the world. In developing countries, however, this number is far lower. In China there are mainly virus-resistant tomato and tobacco. Argentina grows herbicide transgenic soybean, while Mexico has given approval to delayed ripening tomato as well as insect-resistant cotton (Qaim, 1997; James and Krattinger, 1996).

With the exception of the two technologies in China which have been developed by the country itself all transgenic species that have approval in developing countries are products from private companies from industrial countries. Also, over 70 per cent of the field tests with transgenic plants that have meanwhile been conducted in more than 20 different developing countries have been initiated by transnational private companies (de Kathen, 1996). The endogenous modern biotechnology research capacity in southern countries has generally to be regarded as very low. Although comparable information about respective experiences is missing in different developing countries seem to be the most advanced within their region. Especially Sub-Sahara African countries are lacking behind markedly compared to other regions.

All in all a significant discrepancy in biotechnology R&D capacities between industrialised and developing countries can be stated. Biotechnology in developing countries is mainly used in a very applied manner. Apart from a few exceptions the use of modern biotechnology methods in terms of research in developing countries is still at its infancy—which is not surprising keeping in mind the structure of research spending. In 1996 about 4 billion $US had been spent on private and public agrobiotechnology research, the share of developing countries was probably less than 2 per cent.

Effects of Transgenic Plants

Preliminary estimations (James, 1998b) conclude that growing virus-resistant tobacco in China lead to a plus in harvest of 5-7 per cent and a reduction in applying plant protectants by 2- 3 per cent. Growing insect-resistant Bt-cotton in the USA in 1996 lead to savings of 140-280 $US per ha for insecticides and an increase in returns of up to 20 per cent depending on the pressure of infestation. Bt-corn gave an average increase of profit of 9 per cent in the USA in 1996.

In the USA, according to data collection for 1996, herbicide-tolerant soybean lead to significant savings from herbicides, better weed control, higher conservation of soil water, positive effects on harvest and more flexibility in the completion of tillage and cultivation measures. Even more prominent are the expected ecological and economical effects with rape-seed in Canada. Application of herbicides have been reduced to 30 per cent with an average harvest increase of 9 per cent and a significantly higher product quality. Similarly, positive effects have been achieved by growing transgenic potato resistant against potato beetle.

In Mexico a field study about cultivation tests with genetically induced virus-resistant potato has been conducted. Published data illustrates the benefits of the transgenic material over conventional potato varieties especially for family farms with little resources as long as the demonstrated strategies are followed (Qaim, 1998).

Risks of Biotechnology

Just like any other technology there are risks associated with biotechnology. The debate about risks, especially those associated with the application of genetic engineering in agricultural production, is led in a highly emotional manner— and not only in Germany and its North-European neighbour countries. In an unbiased analysis, this is not surprising – especially for many consumers of the North – since it is still in question whom the new developments will benefit and why such technologies should be necessary, giving the fact that there already is overproduction.

Food production, processing and consumption carry an emotional touch and trigger personal feelings that are hardly oriented on whole national economical and/or social dimensions. Furthermore, the use of transgenic crops and animals seems to contradict an environmentally friendly food production and agriculture at first sight.

One has to regard the application of technology in general, and the use of genetic engineering in particular, as a process with a broad spectrum of possible consequences – positive as well as negative. The aims aspired with this technique can also have unwanted side effects.

Safety analysis and impact assessment procedures are, therefore, necessary and generally accepted in the evaluation of genetic engineering programmes. Such analyses tend to be very complicated not only because of the highly complex matter but also because "safety" is no absolute quantity. "Damage" and "risk" are parameters for the assessment of possible unintended outcomes, where "damage" describes the extent of a possible negative outcome and "risk" estimates the likelihood of the negative incident to occur (Käppeli and Schulte, 1998).

The safety issue shall be discussed shortly in the following. Risks are differentiated into those that are technology-inherent and those that are technology-transcendent. Technologyinherent risks are possible negative side-effects on humans, animals or the environment that are directly related to the technology. Technology-transcendent risks include problems that can be triggered by the technology but have its cause in the social, economical and political frame conditions.

Inherent Risks

Especially in the discussion about genetic engineering, concerns have been expressed that the direct change in an organism's genome causes new, unforeseeable and unwarrantable large risks for humans and environment. Since human knowledge is limited the possible existence of such unknown risks cannot be ruled out with absolute certainty neither in the case of genetic engineering nor in the case of any other technology. However, at current knowledge it can be stated that genetically modified plants are not *per se* more dangerous than conventionally bread ones.

Risk assessment can, therefore, not be conducted for genetic engineering in general or biotechnology as a whole, but has to be performed specifically for each single technology product under respective local frame conditions. Here the most important risk areas shall be pointed out briefly and will have to be analysed more closely in each individual case. In modern biotechnology at a level just below genetic engineering, risks are generally rated as being low. The mentioned potential dangers are, therefore, related to commercialisation and consumption of transgenic plants.

The Most Important Environmental Risks of Genetically Modified Crops are:

- Establishment in the wild of the transgenic crop with possible loss of control and biodiversity,
- Gene flow into nature through vertical or horizontal gene-transfer with possible loss of control and biodiversity,
- Development of new viral strains by viral recombination or heterologous encapsidation,
- Breaking of resistance due to selection of pest populations.

The Main Health Risks Mentioned are:

- Formation of new and unintended toxic metabolic compounds,
- Negative effects through antibiotic resistances genes used as markers,
- Rising of new allergens.

First of all, the mentioned risks are the same in developing countries as in industrialised countries. Therefore particular aspects will not be considered in detail, since such discussion would go beyond the scope of this paper. In the case of vertical gene transfer – i.e. the out-crossing of genes of transgenic crops into wild relatives by pollen – the risk can, however, be greater in developing countries than in industrialised countries. This is because of the fact that in developing regions are the centres of genetic diversity of most domesticated crops. While in temperate regions only few wild relatives of agronomically important species occur, there are far more sexually compatible partners in the natural environment of tropical regions, so that a vertical gene-transfer is more likely. Findings from risk studies that are carried out in industrialised countries cannot simply be transferred to countries of the South. This shows clearly that in developing countries great care has to be taken in defining appropriate rules for biosafety as well as in the establishment of effective structures for their implementation. This applies just the same for the area of food-safety.

Such rules have to fulfil mainly two criteria. Firstly a responsible, locally adapted handling of the technology needs to be ensured; secondly limitations for technical advancement should not be too tight. The identification of risk must not automatically mean to relinquish the technology. Risks have to be judged realistically and always be set in relation to the potential benefit of a technology which includes possible benefits to the environment.

The support of establishing and enforcing suitable mechanisms for regulation within partnering countries, also represents an important starting point for developmental cooperation. The development of a regulatory structure is as important as the qualification of specialists and the strengthening of national institutions and structures to ensure the enforcement by operational control and regulatory processes (Virgin and Frederick, 1996). In the context of developmental cooperation with partnering countries the BMZ, GTZ and DSE have put strong emphasis on respective measures to qualify specialists and support institutions.

Transcendent Risks

As mentioned before, technology-transcendent risks are unwanted side effects in a technology that are not caused by the technology itself but by unfavourable frame conditions. Often, aspects of distribution between rich and poor that can be uncovered or enhanced through introduction of a technology, are in the foreground of discussion. The social frame conditions are also of significant importance for the implications of a technology in general and genetic engineering in particular. It has to be clear that technology alone is an inadequate instrument for removing social grievances. Here, other political measures to overcome structural problems are demanded. This is why—in the assessment of biotechnology—technology-transcendent risks should be taken into account and identified early, in order not to counteract developmental aims and perhaps to be able to adequately adjust social frame conditions.

International Imbalances

Internationally, the differences between North and South could increase due to biotechnology, if it is not possible to make technology – including genetic engineering – available and useful for developing countries.

One-sided advances in productivity solely from industrialised countries limit competitiveness of developing countries and create new technological dependencies. Starting points for preventing such a scenario mainly focus on the strengthening of national agricultural research systems with clear priority setting, a national and international efficient cooperation between private and public research, and the definition of suitable intellectual property rights (IPR) regulations.

National Imbalances

An aggravation of differences could also arise within developing countries if, for example, large companies use genetically improved seed material, not available to rural family farms. Although genetic engineering can principally be regarded as being independent of business size from the technical point of view, in rural family farms the frame conditions for acceptance of technologies are often less favourable. For example poor households generally have less access to information and extension services that are crucial for the adoption of technical innovations. Furthermore temporary financial limitations at the time of sowing can also aggravate the acceptance of technologies by small holdings with little resources even if the technology is generally profitable. Adequate access to agricultural extension service and credits as well as a well functioning seed market by which the technology reaches farmers are important prerequisites, so that unintended aspects of distribution within a country can be prevented. National technology politics must also not be restricted to research but has to explicitly include the area of technology distribution and application (Brenner, 1996; Qaim, 1998b).

Raw Material Substitutions

Concerns that don't have anything to do with the use of the technology in developing countries itself and still affect these strongly in a negative sense, are those of possible raw material substitutions by biotechnology and genetic engineering processes. It is foreseeable that certain raw materials that so far could only be obtained from tropical plants will also be obtainable in industrialised countries in the future through biotechnology.

For example, certain valuable fatty acids can so far only be extracted from cocoa butter or coconut oil. A genetic modification of oil crops of temperate regions can perhaps uncouple the fatty acid production from the originally producing plants in the future (Franzen *et al.*, 1995; TAB, 1995). Other substitutions of typical raw materials from developing countries through biotechnology measures are expected in the area of herbal and aromatic substances. Research is, for example, conducted to produce natural vanilla aroma in the lab by means of cell and tissue culture. Such raw material substitutions can drastically influence the trade fluxes of tropical products and, therefore, reduce export chances of producing countries.

For developing countries, whose economy often depends strongly on respective raw material exports, that could have serious consequences on the whole national economy. From a developmental policy point of view, this would be highly unwanted and these problems should not be underestimated.

However, it has to be clear that possible raw material substitutions by new technical processes are not a special feature of biotechnology.

The demand for caoutchouc for rubber production decreased in the course of technological development just like the demand for copper as electrical conductors. The dependence on only one or a few raw material export products is in the long run always connected with a high risk, since economic mechanisms – with or without biotechnology aim to substitute more expensive raw materials by cheaper substitutes. Instead of interfering by slowing down innovative developments, it is more sensible to work towards a diversification of the economy in potentially affected countries (Leisinger, 1995). This could be supported by international and bilateral developmental aid efforts.

Contribution of biotechnology and genetic engineering to a sustainable development The reduction of hunger and poverty is a basic prerequisite for a sustainable development. A threat for natural resources often originates from poverty and is due to overuse of resources forced upon by poverty. Hereby the relation between poverty and biotechnology and genetic engineering is of a complex nature.

Therefore, in analysis and evaluation of biotechnology and genetic engineering one has to regard especially the potential of this technology for low income groups in developing countries, and a comprehensive impact assessment including economical and cultural parameters has to be conducted.

A sustainable development according to the Agenda 21 is given, if it is economically efficient, socially equitable and responsible as well as ecologically acceptable (UNEP, 1994). Economic efficiency supports the assumption that resources are optimally employed and neither overstrained nor inadequately used. It is quite obvious that due to the very fast advances in cell and molecular biology working materials (mainly seed material from transgenic plants), increasingly obtained with the aid of biotechnology and genetic engineering, offer potential for the more efficient use of natural resources like genetic material, water, soil and nutrients.

However, one has to keep in mind that an optimal allocation of resources in the sense of a sustainable development is only possible, when the price of the resource reflects more or less all costs incurring in the use of this resource. In order to achieve this ecological as well as social costs need to be internalised. The socio-economic restrictions in the use of genetically modified organisms as well as the ecological risks of the technologies have already been discussed.

The ecological risks imposed by the application of biotechnology and genetic engineering hold the weakest position in respect to sustainability, whereas the socio-economic restrictions are not specifically technologyinherent.

Regarding the total extent of possible applications for genetic engineering (especially in the area of food security and environmental applications) it is today undoubted by many specialists that genetic engineering has the potential to contribute positively to a sustainable development. The application of technology enhances economic efficiency in the use of resources, has the potential to be socially equitable and ecologically responsible under appropriate frame conditions. Therefore, biotechnology and genetic engineering are accepted technologies in the Agenda 21 for the achievement of a sustainable development.

However, genetic engineering – as any other technology— is demanded to be environmentally safe and sound and tied up in an institutional framework that supports sustainable development. Therefore a respective education and advanced training for the users of genetic engineering is rightfully demanded (UNEP, 1994).

In several developing countries the priority in biotechnology R&D lies in the agronomy sector. This is partly due to the necessity to ensure the feeding of a growing population at constant or rather decreasing resources of arable land, water and energy in the short and in the long term—on the other hand, due to the necessity to increase the productivity of agricultural export goods, that are of prime economic importance in many developing countries.

However, in order to use especially the comparative benefits of biotechnology and genetic engineering in respect to the sustainability of development, the applications of this technology have to be extended also to other areas in the industrially expanding states (especially of Newly Industrialised Countries). Waste disposal and avoidance needs to be emphasised as well as applications in the reduction of environmental pollution and the rehabilitation of damages to ecology.

A special relation, with regard to the sustainable development, exists between the application of biotechnology and genetic engineering procedures and the conservation or degradation of genetic resources.

In medium-term, further developments in agriculture by biotechnology and genetic engineering, pharmaceutical industry, environmental technology and for new materials, will increase the demand for (plant-) genetic resources (von Braun and Virchow. 1998).

Increasing the significance of safe conservation processes (*in situ* and *ex situ*) and more efficient access regulations for genetic resources is a prerequisite for demand-driven biotechnology-assisted propagation and breeding activities (Virchow, 1997c).

With the further development of genetic engineering and increasing recording (and conservation) of genetic resources as well as a possible synthesis of genetic material and its further recombination, the significance of – and the demand for protected genetic resources can possibly decrease in the long term. This could be true, if the *in vitro* reproduction of genetic information would ever be cheaper than currently practised (*in situ* and *in vitro*) conservation of the genetic resources.

The substitution of natural genes is so far only speculative. In the mid-term, the use of genetic resources seems to be cheaper, facing the possible increase in efficiency in its conservation (von Braun and Virchow, 1995).

Role of Private and Public Sector

The private sector dominates the worldwide research in the field of agricultural biotechnology. Although exact data about international research expenses is not available, it is estimated that private companies account for about 75 per cent in this area worldwide, with increasing tendency.

Few transnational companies from industrialised countries dominate biotechnology research and the degree of concentration on the markets for the respective technology products is growing.

Alone in the last three years, mergers, company take-overs and alliances in the area of agrobiotechnology account for about 20 billion $US. However, the worldwide turnover of transgenic seed material only amounted to an estimated 0.5 billion $US in 1996. This clearly shows that especially long-term interests of companies are causes for the process of concentration.

In order to secure shares for the fast growing markets of the future and to be competitive, there are three main strategic reasons for merger and take-over tendency: Firstly, the portfolio of patents to be available can be broadened in this way. Secondly, a critical financial volume is created in order to cover for long-term and high-risk areas of research and thirdly, vertical integration occurs in order to be able to collect greater market margins.

The question remains, what implications this private dominance in research in the North has for the application of biotechnology in developing countries. Furthermore the biotechnology developments also have effects on public research institutions that partly have to redefine their role.

Private Sector

Looking at the different products from biotechnology and genetic engineering of transnational companies already commercially available today or will be in the near future, it is conspicuous that a series of these technologies could also be suitable for developing countries (for example transgenic resistances against biotic stresses in maize, cotton, potato, etc.). However, there are only sporadic applications of these products within the countries of the South so far. Obviously, reasons exist that so far still hamper a North-South transfer of technology:

- A seed-technology that has been developed in the North has to be adjusted to local varieties at first before it can be applied at other locations. Therefore, a technology adjustment is needed that should generally be conducted within the receiving country. In most developing countries, there is a lack of specific know-how and adequate R&D infrastructure in order to supply partners for a North-South technology transfer. Transnational companies that have hardly been active in the seed sector of developing countries so far, lack own experience to conduct such technology transfer.
- Seed markets are poorly developed in many developing countries and often characterised by inefficient governmental seed producing and distributing programmes. In addition – or in part also due to this—especially the great number of rural family businesses often reproduce their own seed material. The positioning of a seed technology within such markets primarily requires the creation of new efficient distributing institutions. This environment increases the risk for commercial failure of private companies.
- Intellectual property rights (IPR) for plant varieties don't exist in many developing countries or they are unattractive for private innovators. It is often argued that patent regulations aggravate the access to

technology for developing countries. In order to get access to privately developed technology, it is in contrast important to establish effective mechanism for the protection of intellectual property.

- In many developing countries issues of biosafety are not clearly regulated or there is a lack of experts and infrastructure/institutions as prerequisites to put the legal regulations efficiently into practise. Since – as explained above – the genetic centres of many domesticated crops are in the tropics, it is absolutely necessary, reflecting the threat for loss of biodiversity, to firstly establish operational regulatory mechanisms.

Missing or cumbersome institutional competence hinders an efficient biological safety control, lengthens the approval process for biotechnology products and retards private interest.

When these obstacles can be eliminated, one can count on a larger interest of the private biotech and seed sector in the markets of developing countries. In the passed 20 years, the quantitative use of seed material stagnated in industrialised countries or even decreased somewhat. Quite in contrast, the use of seed material increased in most developing countries, particularly in Asia and Africa. So these are the growing markets for seed technology.

Therefore, it can be assumed that the private interest in developing countries will rise in the long term. Naturally this will come essentially from private business interest and not from idealism and sympathetic considerations. Thus the private industry is hardly going to worry explicitly about the special interests of rural family businesses. To make new technologies available for these groups is an important task to be taken up by national programmes. Under suitable institutional frame conditions within developing countries it is quite possible to achieve all-round advantages (win-win situation). To renounce the contribution from private industries considering the great advance in research of such companies compared to the public sector, must appear to be utopia. In order to exhaust the potentials of the private economy the well-directed strengthening of national agricultural research systems in terms of technology adjustment as well as the creation of suitable institutional frame conditions for technology acquisition, spreading and use in developing countries should enjoy high political priority.

It is not a lack of basic technology but a lack of capacities and mechanisms to adapt the already developed biotechnology procedures and products to the local requirements and to make them available to the users. That is, at present, the limiting factor for the efficient distribution of biotechnology in agriculture of developing countries.

There are already a few internationally operating initiatives, acquiring and providing private biotechnology products and procedures, meeting the demand of potential applicants in developing countries. The problems addressed above are explicitly considered in the projects undertaken. Two of the initiatives are to be introduced briefly:

- The *International Service for the Acquisition of Agri-biotech Applications* (ISAAA) is an international non-profit organisation that provides institutional assistance in North-South technology transfer. ISAAA was founded in 1991. Apart from the identification of target groups in selected partner countries and suitable technologies, their main concern is the support in mediating contractual arrangements of the transfer as well as the adjustment and application of the technology in a developing country. Thereby the necessary measures for specialist qualification and for the development of institutional frame conditions are considered. The strategy is to create an infrastructure for technology development and distribution with the aid of pilot projects. Target groups are mainly resource poor family businesses in Africa, Asia and Latin-America.
- The Project *Agricultural Biotechnology for Sustainable Productivity* (ABSP) by USAID also started in 1991. ABSP is a Network of scientists from the public and

private sector that conduct applied research in partnership, focusing on tissue culture and genetic engineering. The aim is a transfer of technologies followed by the establishment of structures for product development within developing countries. Indonesia, Kenya and Egypt are in the focus.

Public Sector

From the background of the private research dominance in the area of agricultural biotechnology the question arises, which role public research at universities or national and international centres could have in this. Within the centres of the *Consultative Group on International Agricultural Research* (CGIAR), which initiated the impulse for the Green Revolution in the 1960's, about 24 million $US was invested in biotechnology (Manicad and Lehmann, 1997).

The international donor community to the CGIAR invested a total of about 400 million $US in agrobiotechnology through bilateral and multilateral programmes during the past ten years (Brenner, 1996). According to conservative estimations, this nearly equals the yearly agrobiotech research budget of the US company Monsanto.

The different dimensions of research investment clearly shows, that there cannot be an immediate competition between the private and the public sector for biotechnology markets. Given their research advance and a better financial and technical equipment, private companies have crucial comparative advantages and can develop comparable biotech-products far more efficiently.

Nevertheless, there are a number of niches where public research institutes have to give important contributions. An advantage of genetic engineering lays in the ability to adjust plant varieties to their agroecological demands of special regions far better than this could be achieved by using traditional breeding programmes. Even though ertain technical developments of the North can often also be suitable for the countries of the South, this does not cover the whole spectrum of defined demands and necessary technologies in developing countries.

Partly the agronomical requirements under tropical and subtropical circumstances are different to those in temperate climates. Apart from the fact that other plant species are being cultivated, also the climatic and soil biological demands of the plants may differ.

Speciality technologies that could be highly interesting in developing countries, but being hardly or not at all an objective in industrialised countries, private research interest cannot be expected in the short or medium term. This includes for example biotechnology research on manioc, sorghum, sweet potato and other regional staple food crops (*orphan commodities*). Furthermore, the market potential for traits like drought and salt tolerance or the breeding for better adaptability to those micronutrients that can be fixed in tropical soils, is probably too small and the marketing risk too high in order to be of interest to private research. It is the duty of public research to identify these niches and set clear research priorities to them. One example for an international network of public institutes for research on manioc is the *Cassava Biotechnology Network* (CBN) which is coordinated by the *Centro Internacional de Agricultura Tropical* (CIAT) in Columbia.

Similar initiatives in the area of rice research are the *International Programme on Rice Biotechnology* of the Rockefeller foundation and the Asian Rice Biotechnology Network.

Cooperation between Private and Public Research

Even though the scope of function for private and public research can roughly be defined with respect to the development of technology products, it remains necessary to intensify the cooperation between these sectors. Especially basic knowledge in biotechnology (e.g. marker-genes and promotors, technologies for transformation/regeneration, etc.) is applicable for a vast spectrum of end technologies in product development. It would be very inefficient not to apply available know-how in the private and the public sector at the same time (James, 1997b). Apart from that, duplicated research leads to legal problems.

The patent holder of a promotor sequence which is used in genetic transformation can for example inhibit commercialisation of a technology, even if the product is entirely new and unrelated. With increasing patent-mediated protection, the situation will become more and more complicated. Even in a transgenic plant, where only a single gene has been changed, eight different patents or similar *Intellectual Property Rights* (IPR) can easily be involved (Sehgal, 1996). As expected, the private sector is far ahead of the public sector in terms of patenting.

Three of the world's largest agrobiotechnology companies (Monsanto, Novartis, AgrEvo) alone hold about 70 per cent of all patents in the sector of plant biotechnology. Without having an interesting patent portfolio themselves, it will be hard for public organisations to negotiate for access to the necessary basic technologies in order to conduct independent biotechnology research.

On the other hand, private companies need access to genetic resources for crossing, isolation of certain genes and production of transgenic material harbouring agronomically interesting genes. Since the *Convention on Biological Diversity* (CBD) came into force in 1993, such genetic resources are covered by the sovereignty of the countries of origin - for the main part of interesting materials these are developing countries. These complexities demand for new forms of cooperation between the public and private sector. The CGIAR-system reacted to the changed frame conditions by establishing a *Private Sector Committee* (PSC) in 1995. The PSC consults on potential mechanisms of interaction between CGIAR, national agroresearch systems and private sector. The setting of IPR regulations within *international private-public joint ventures* acceptable for all sides is a challenge for the future.

Policy Framework for the Application of Biotechnology and Genetic Engineering

Intellectual Property Rights: There exists a multitude of options for safeguarding intellectual property for the promotion of further investments in biotechnology and genetic engineering R&D. Representative lawful regulations about *Intellectual Property Rights* (IPR) in the area of genetic engineering and plant genetic resources still take place on national level in industrialised as well as developing countries.

The safeguard of IPR for agricultural innovations has been handled quite differently within developing countries but is generally considered soft.

In Brazil, for example, plant and animal species as well as biological processes for their production are excluded from patenting. Biotechnology patents are confined to special commercial applications of microorganisms. Plants and animals cannot be patented. The patent law in India from 1970 excludes agricultural products, the Chinese patent law from 1993 alike (van Wijk *et al.*, 1993).

The agreement about *Trade Related Aspects of Intellectual Property Rights* (TRIPs), recently accepted within the scope of the GATT negotiations, regulates that "...plants and animals other than microorganisms and essentially biological processes for the production of plants and animals may be excluded from protection".

Furthermore the TRIPs regulations have to be put into practice within 5 years (i) exceptional cases 10 years) after completion of the agreement in developing countries, out of which about 70 have signed the agreement. It can be expected that in the future, IPRs on biological products in developing countries will still lag behind those in industrialised countries (von Braun and Virchow, 1998).

The national as well as international protection of intellectual property through TRIPs in the scope of the WTO that will come into action in all signatory countries, has implications in three national as well as institutional meaningful areas:

- Firstly the development of innovations in biotechnology and genetic engineering will depend on the national degree of protection. Private R&D will develop only, where profit expectations are comparable to investment risks. A country can only allow itself to grant low or no protection of intellectual property, if it regards R&D in the area of biotechnology and genetic engineering as public good and can compensate the lack of private activities by public investments.
- Furthermore, the use of internationally available technologies in R&D of National Agricultural Research Systems (NARS) is limited by the already existing rights of protection.
- Finally, all existing and planned mechanisms for protection of intellectual property must take into account and enable technology transfer to the user, especially farmers.

The demand for protection of intellectual property has increased with the introduction of genetic engineering processes since higher and more risky investments in R&D have to amortise itself, the intensity of competition between respective companies increased and strategic competitive advantages can provide a significant advantage. Mainly process patents (in contrast to product patents) can concern a whole set of possible applications.

Defence and recovery of intellectual property can, therefore, enhance the development of the plant breeding sector in developing countries and the sustainability in the application of biotechnology and genetic engineering processes in national breeding programmes.

Potential winners in the enhancement of proprietary rights in the area of plant breeding are developing countries with high technology potential as well as those with a large agrobiodiversity (Virchow, 1998). The disadvantages of an enhanced protection of ownership are mainly the costs for organisational and institutional development and maintenance of the defence and recovery system. Furthermore, many developing countries—being net importers of technology – will have to face increased costs for the use of introduced technologies. Developing countries with native seed production that uses foreign technologies that are hardly protected so far, will probably have to cope with a loss of wealth because of increased regulation of ownership protection. On the other hand, if national IPRs are not introduced, costs because of the fact that certain advanced technologies are not available to the country, or coming in with some time delay, must not be neglected (Lesser, 1997).

There is a whole set of different IPR mechanisms out of which the two most important ones in the area of plant varieties shall be introduced briefly.

- *Patents:* Patents grant an inventor or developer of a new plant variety a temporally restricted monopoly for its commercialisation, i.e. the direct marketing of the variety to outsiders, but also its propagation by the farmer is prohibited or requires licensing. The further use of a variety for research and breeding by others is not strictly regulated. At the latest commercialisation of a successor variety requires also license approval by the original patent holder.
- *Plant Breeders' Rights (PBRs):* PBRs is an original protection mechanism for new varieties of crop plants in the scope of the UPOV agreement. Similar to patents, it provides a frame of protection to the inventor of a variety that restricts commercialisation through others. The use of a new variety for further research and breeding purposes by others is explicitly granted.

The protection of copyright through PBRs is generally regarded to be weaker than patents because of exceptional regulations. The TRIPs agreement principally provides for patenting of plant varieties that must be warranted by the signatory countries.

Alternatively, a patent can be omitted if an effective *sui generis* system for the protection of copyright of new varieties is present. According to this verbalisation the PBRs system should be accepted in one of the UPOV agreements. Many developing countries that have signed the TRIPs agreement so far, neither provide for patents nor are they members of UPOV so that there is an institutional call for action for the signatory countries in order to meet the GATT regulations.

Another aspect gaining importance in biotechnology development is the fact that there is an increasing demand not only to grant copyright protection for plant varieties as the final product but also for biological processes and single genes. Such demands cannot be covered with mechanisms of PBRs. Developing countries face the challenge to identify individual regulations that allow access to international technology, support national innovation activities and guarantee availability of technology products for resource poor users at the same time.

Bio and Food-safety

The use of biotechnology and genetic engineering in agriculture as well as food technology can only make a positive contribution to sustainable development in the sense of UNCED, if the technology-inherent risks are detected and controlled. The safety of applying biotechnology and genetic engineering is a central point of the *Convention on Biological Diversity* (CBD) as well as in the second "Conference of Parties" taking place after UNCED.

A *Protocol on Biosafety* was demanded there (UNEP, 1995). Biosafety and Food-safety—defined as the policies and measures assigned to enforce the environmentally safe applications of biotechnology and genetic engineering in agriculture and food processing—gain national as well as international importance in this context (Persley *et al.*, 1993).

General principles and standards should regulate the introduction of transgenic organisms into the environment and in the area of nutrition and serve politicians or scientists as a baseline for their work.

Unfortunately, several forces act against the compliance with a recognised code of conduct. On the one hand, the awareness for the potential risks resulting from using biotechnology and genetic engineering processes is very heterogeneous on an individual as well as on the national level. The differences are partly due to the different intensity of risk assessment but also due to the uneven level of information of the concerned parties.

On the other hand, since countries are differently advanced in their technology, the susceptibility to accidents in biotechnology and genetic engineering plants is not evenly distributed (von Braun and Virchow, 1996).

Finally, the pressure to increase food production as well as irresponsible profit-making may lead to an environment, where risks are put in perspective or neglected and safety instructions for biotechnology and genetic engineering processes are ignored.

The wording, legal passing and implementation of regulations alone will not be sufficient in order to grant a safe handling of biotechnology and genetic engineering, especially in developing countries with a low technological standard. Mainly assistance in capacity building for the effective implementation of regulating instruments is required (Frederick and Virgin, 1996).

Among others, genetic engineering has the potential to improve food supply and health care, increase efficiency in waste recycling, detoxify hazardous compounds and increase forestry production.

Nevertheless, it is often argued that especially genetic engineering is unsuitable for developing countries since it is a form of non-adapted high-technology (Buntzel, 1997; Lange, 1997). Although genetic engineering research is certainly a part of high-technology this does not hold true for the agricultural area with the technology product finally applied by the farmer—improved seed material.

Seed material is an arbitrarily dividable resource, so that even rural family businesses, that were so far often disadvantaged when technical innovations were introduced, could benefit from the advances in biotechnology and genetic engineering.

The argument that—before genetic engineering is applied—all potentials of conventional cultivation methods should be exhausted (Wolpers, 1996), is basically right. However, it does not sufficiently take into account that the final utility of biotechnology and genetic engineering already equals or even exceeds the final utility of conventional technologies in many cases today.

Furthermore, within agricultural research one should think in long term. Because of the time needed, it could be fatal not to start developing new ideas only when no further progress in productivity is possible through conventional processes. In addition, it can not be ruled out that genetically engineered agricultural products, especially for resource-poor farmers, have the advantage of being far easier to handle—and therefore be more promising—than conventional cultivation measures.

Many of the technologies integrated in the seed material do not require complementary inputs so that they would be easy to integrate in existent cultivation practice. All the same genetic engineering should not be regarded as a replacement but as a promising enlargement of conventional technology instruments that give an important contribution to food security and to agricultural growth.

For a global food security of a constantly growing world population, it is of course important not to neglect increases in food production in favoured regions of industrialised countries. Even so we must work towards increased application of biotechnology processes and products in the mostly agriculturally marked developing countries.

In order to make the potentials of biotechnology available for everybody and to prevent an increase of the technology gap between North and South, the capacity for biotechnology R&D should be supported and expanded.

In many developing countries, biotechnology is already discussed on a political level. Nevertheless, there is often a lack for a clear priority setting and an adequate resource allocation. In addition, the mandates within the national agricultural research systems are not always clearly set apart, so that an efficient use of the anyway limited resources is inhibited.

Developing countries should, therefore, first formulate a locally adapted and demand-oriented biotechnology policy that is tailored to the comparative advantages of each country. This does, however, not mean that each country must be able to conduct its own basic research in biotechnology and genetic engineering. Especially for poorer countries, this aim could lead to wrongly set priorities.

Rather increased attention should be on potential North-South and South-South technology transfer. For this, it is a prerequisite that a minimum of know-how and institutional infrastructure in the area of applied biotechnology research, product development and processing is present in the country.

Those countries of the South that already have some experience with biotechnology could play an important role for spill-over into neighbouring countries and, therefore, support regional technology development.

A specific cooperation can develop between countries with genetic resources and technologically advanced countries that depend on the access to genetic material. In order to shape this efficiently, the developing countries rich in biodiversity should try to create capacity in the systemic gathering and preparation of genetic resources in order not to be just an exporter of genetic material, but to harvest potentials of appreciation themselves.

The private sector is only active where profits are likely in the short or medium term, market structures exist or are easy to establish and investments are protected.

Since these prerequisites are not given in many developing countries, the private biotechnology and genetic engineering sector of industrialised countries, which can be regarded as the most important instrument for an efficient use of natural resources, often does not find appropriate incentives to direct its R&D and production activity towards the problems of developing countries.

Private commitment from industrialised countries in the near future can only be expected, partially and punctually, in the commercial seed sector of developing countries. These deficits must widely be compensated through publicly financed R&D efforts in order to enable the use of biotechnology and genetic engineering in agriculture.

In principal, this also holds true even for conventional agricultural technology. New forms of cooperation with the private sector—, e.g. recourse on private basic research results for payment—are possibly to be developed in order to realise genetic engineering transfer in developing countries.

The *modus operandi* of ISAAA as a thinkable cooperation model in the North-South and South-South transfer has been explained above. In this, the national agricultural research systems in developing countries deserve top priority, they have to be enabled to put internationally available research results into practice and make it available to the target group of agricultural producers.

The contribution of private genetic engineering to global sustainable development depends on the readiness for cooperation on three levels:

- Cooperation between the private owners of technology

and public owners of genetic material for a sustainable use of genetic resources;

- Cooperation between the genetic engineering industry and public research facilities to the benefit of global food safety;
- Cooperation on the international level for the elaboration and surveillance of safety standards and IPR regulations.

But even in constructive cooperation there will be losers: technology and resource poor countries will neither profit sustainability from the increasing significance of genetic engineering nor from the increasing demand for genetic resources.

For these countries especially South-South co-operations will be of great relevance. Hereby technological spill-over from regionally advanced countries with similar agroecological conditions could occur.

The support of South-South biotechnology transfer—including the ridding from institutional problems in implementation—is an important area for technical cooperation.

The realisation of the potentials of biotechnology and genetic engineering in developing countries and the minimisation of costs and negative side effects is an aim, which will not engage itself. Intelligent politics on national and international level is required, that promotes, accompanies and leads technology development into desired directions.

Information in-time about technological implications will become a key issue in the political decision process. However, since there are only few applications of modern biotechnology in developing countries so far, such an information baseline is more or less absent.

Experiences with conventional technologies, as for example in the Green Revolution, give important hints. One must not miss the fact, that biotechnology and genetic engineering in many respects has different and new characteristics, so that it can also be counted on that the consequences will be different too.

The area of technology impact assessment tries to examine the consequences of new technologies systemically. Hereby mainly qualitative statements are made that refer to a technology in total.

Meanwhile, the issue is not the constitutional decision whether biotechnology and genetic engineering should be applied and supported in developing countries, but rather which intensity and direction it should take.

Therefore, detailed information about potential consequences of specific technologies are needed that explicitly account for the mutual dynamics of technological and social development (Qaim and von Braun, 1998).

Quantitative economical *ex ante* methods can give an important contribution for widespread technology assessment within case studies in developing countries. Currently, there is a substantial research deficit and unfortunately only very limited readiness in the North and in the South to supply the necessary public means for *ex ante* technology assessment.

Industrial Biotechnology

Achievements, Prospects and Perceptions: 'White' biotechnology or industrial and environmental biotechnology is a broad and expanding that includes making enzymes with a variety of industrial uses that include the manufacture of bioplastics and biofuels and using microorganisms and plants for the treatment of wastes and abatement of pollution, a process known as bioremediation.

It may seem to be occupying a small niche now, but its products or processes will become as ubiquitous as those of the chemical industry today because of the benefits they pose to the environment and the premium attached to this environment-friendliness for the companies that undertake them.

Bacterial bioleaching, i.e. using specific bacteria to extract metals from ores or mine wastes sprayed with water, is a growing sector of the mining industry and several developing countries are already playing a key role in this area.

Other research efforts are scanning the microbial diversity of various environments (e.g. the oceans and seas) and deciphering their genetic information aimed at isolating microorganisms that could be used in the manufacture of drugs, enzymes and a wide range of bioactive compounds, as well as in bioremediation processes. As for materials and fuels that are not derived from petrochemical processes, researches are also underway for processes to improve and enhance the bioremediation of water, soils and ecosystems at large, and minimise the use of fossil-fuel energy. All these forms of 'white' biotechnology are poised to increase the eld's positive social acceptance.

7

New Marine Biotechnology

Options and Opportunities

Life originated in the sea and has sustained itself to the present day. The world's oceans comprise the largest part of the biosphere and contain the most ancient and diverse forms of life. The marine biotopes contain an unmatched metabolic and organismal diversity.

Over billions of years marine microorganisms have moulded the global climate and structured the atmosphere. The quantum leap of our understanding of life during the last one decade through the advent of genomics and bioinformatics at an organismal, cellular, or genetic level has opened new perspectives.

The oceans offer abundant resources for research and development, yet the potential of this domain as the basis for new biotechnologies remains largely unexplored. Indeed, the vast majority of marine organisms have yet to be identified. Even for known organisms, there is insufficient knowledge to permit their intelligent management and application. This vast sea of life from microscopic bacteria to towering tubeworms is considered to be a largely unopened treasure chest that, with further exploration, will yield new products and processes of benefit to humankind as well as the environment. Our challenge as a nation is to develop the biological technologies needed to identify sources of ecological stress to develop strategies to protect and restore marine and coastal resources.

Concept of Marine Biotechnology

Biotechnology is defined as the industrial use of living organisms or biological techniques developed through basic research. In recent decades, however, the term "biotechnology" has been more often applied to cutting-edge molecular or genomic biological applications where molecular or genetic material is manipulated to achieve a desired goal (Bunker *et al.*, 1994). Targeted drug discovery, DNA 'fingerprinting', gene amplification and sequencing and the use of recombinant DNA techniques to genetically 'engineer' organisms, are among the scientific endeavours that fall under this interpretation of the term.

Broadly defined, marine biotechnology is "the exploration of the capabilities of marine organisms at the whole, cell, or molecular level, to provide solutions to today's problems, with the use of technology to advance the understanding and accessibility of marine biological material" (Lee and Burrill, 1994). Simply, marine biotechnology seeks to develop goods and services from the organisms and processes of the ocean. For example, a compound from coral is used as an inflammatory drug, as part of a larger quest for new medicines. Other scientific investigations are seeking ways to detect environmental contaminants, discover new industrial compounds and assure quality of seafood.

Importance of Marine Biotechnology

The oceans encompass about 71 per cent of the surface of our planet, but over 99 per cent of the biosphere, and they represent the greatest extremes of temperature, light and pressure encountered by life (DeLong, 1992). Adaptation to these harsh environments has led to a rich marine bio- and genetic diversity with potential biotechnological applications related to drug discovery, environmental remediation, increasing seafood supply and safety and developing new resources.

Oceanic organisms are of enormous scientific interest, for two major reasons (Zilinskas *et al.*, 1994). First, they constitute a major share of the Earth's biological resources. Second, marine organisms often possess unique structures, metabolic pathways, reproductive systems and sensory and defence mechanisms because they have adapted to extreme environments ranging from the cold polar seas at -2° C to the great pressures of the ocean floor, where hydrothermal fluids spew forth.

Most major classes of the Earth's organisms are primarily or exclusively marine, so the oceans represent a source of unique genetic information. Additional support of research in key areas of marine biotechnology will generate both new fundamental knowledge and advanced technologies for producing new pharmaceuticals, biomaterials and other products; developing and improving bioremediation and bioprocessing; enhancing cultivation of aquatic species; and expanding understanding of biological processes in the oceans and their role in global change. However, it is crucial that any exploitation of marine resources be done in a manner that is sustainable and does not damage the ecosystem.

Major Applications and Benefits

Marine biotechnology has demonstrated its potential across a broad spectrum of applications that range from biomedicine to the environment. Nevertheless, despite noteworthy successes and the inherent promise of the ocean's vast biological and chemical diversity, marine biotechnology has not yet matured into an economically significant field (US-EC Task Force on Biotechnology Research, 1992). Fundamental knowledge is lacking in areas that are pivotal to the commercialisation of biomedical products and to the commercial application of biotechnology to solve marine environmental problems, such as pollution, ecosystem disease, and harmful algal blooms. The major applications and benefits of marine biotechnology are as follows:

Novel Drugs and Marine Derived Compounds: Drug discovery represents one of the most promising and highly visible outcomes of marine biotechnology.

Biochemicals produced by marine invertebrates, algae and bacteria, are very different than those from related terrestrial organisms and thus offer great potential as new classes of medicines (Martin *et al.*, 1993). To date, examples of marine-derived drugs include an antibiotic from fungi, two closely related compounds from a sponge that treat cancer and the herpes virus, and a neurotoxin from a snail that has painkiller properties making it 10,000 times more potent than morphine without the side effects.

However, there are several more marine derived compounds currently in clinical trials and it is likely that many more will advance to the clinic as more scientists look to the sea for these biotechnological uses.

In addition to new medicines, other uses for marine-derived compounds include cosmetics (algae, crustacean and sea fan compounds), nutritional supplements (algae and fish compounds), artificial bone (corals) and industrial applications (fluorescent compounds from jellyfish and novel glues from mussels) (Tiedje, 1994).

Marine microorganisms are also a source of new genes, the exploitation of which is likely to lead to the discovery of new drugs and targets (Martin *et al.*, 1993). Secondary metabolites produced by marine bacteria and invertebrates have yielded pharmaceutical products such as novel anti-inflammatory agents (pseudopterosins, topsentins, scytonemin and manoalide), anti-cancer agents (bryostatins, discodermolide, eleutherobin and sarcodictyin) and antibiotics (marinone). Anti-oxidant peptides have been isolated from protease digests of prawn muscle, and the mycosporine amino acid (MAA) precursor, 4-deoxy-gadusol, has been isolated from seaweeds. These have applications as food additives and in cosmetics. To date, exploitation of natural agents from the sea has been hindered by problems with limited or sporadic distribution and production. Much more research must be conducted to determine what seasonal factors and life cycle or reproduction states are linked with natural production of an agent. Knowledge of these factors will be important in developing methods for producing selected metabolites, either from whole organisms or *in vitro* from cell or tissue cultures of plants and animals.

Enzymes and Biocatalysts

Enzymes produced by marine bacteria are important in biotechnology due to their range of unusual properties. Some are salt-resistant, a characteristic that is often advantageous in industrial processes. The extracellular proteases are of particular importance and can be used in detergents and industrial cleaning applications, such as in cleaning reverse-osmosis membranes.

Vibrio species have been found to produce a variety of extracellular proteases (Committee on Life Sciences and Health, 1992). *Vibrio alginolyticus* produces six proteases, including an unusual detergent-resistant, alkaline serine exoprotease. This marine bacterium also produces collagenase, an enzyme with a variety of industrial and commercial applications, including the dispersion of cells in tissue culture studies. Other research has demonstrated the presence in algae of unique haloperoxidases. These enzymes could become valuable products, because halogenation is an important process in the chemical industry. Japanese researchers have developed methods to induce a marine alga to produce large amounts of the enzyme superoxide dismutase, which is used in enormous quantities for a range of medical, cosmetic and food applications.

Thermostable enzymes offer distinct advantages, many

still to be discovered, in research and industrial processes (Wackett, 1994). Thermostable DNA-modifying enzymes, such as polymerases, ligases, and restriction endonucleases, already have important research and industrial applications.

Hot springs in Yellowstone National Park provided the first archaeon *(Thermus aquaticus)* from which thermostable DNA polymerases were isolated. These novel enzymes became the basis for the polymerase chain reaction (PCR), a useful technique for studying genetic material.

Most enzymes involved in the primary metabolic pathways of thermophilic bacteria and archaea are dramatically more thermostable than are their counterparts living at moderate temperatures.

Expanded study of enzymes from thermophilic marine microorganisms will contribute to the understanding of mechanisms of enzyme thermostability and should enable the identification of enzymes suitable for industrial applications as well as modification of enzymes to enhance thermostability.

Nature's catalysts function highly selectively to accelerate the rate of chemical reactions in biological systems. For many substrates, enzyme-catalysed reactions exhibit rates that are many orders of magnitude faster than those of uncatalysed reactions. At present, about 50 enzymes are used in industry. Most of these industrial uses involve breaking down large molecules into simpler ones. Examples include the proteases in laundry detergents and the hydrolases used in starch processing.

Composite Materials and Biopolymers

Macroalgae (seaweeds) have been used in Asia, for more then two thousand years as a subsidiary food and have also been used extensively in medicine. Macroalgae make use of different photosynthesising pigments that divide them roughly into three groups: the brown, green, and red algae, in contrast to the terrestrial environment which has only the green plants.

From a biotechnological point of view, the biochemical diversity in seaweeds provides a potentially enormous variety of products, some of them of current biotechnological use.

For example, polysaccharides from certain red algae (carrageenans and agars) or brown algae (alginates) are still largely used as jellifying or thickening agents in the food industry, in cosmetics or even in building materials (National Research Council, 1989). More recently, seaweeds have been used in aquaculture diets for urchin, abalone and fish as an alternative food source, due to the scarcity of fish meal and fish oil and the rise in costs in the feed industry. These seaweeds used in the feed contain all the essential amino acids, PUFAs, vitamins and minerals, and include a small portion of the protein as well.

Seaweeds, in general, are high in essential minerals and vitamins and have been used in the human diet as a health food or for therapeutic uses.

The brown seaweed, *Fucus serratus* is useful in thalassotherapy (Weiss, 1994). It has been proved that compounds such as laminarin and fucoidans are known to have an anti-tumoral action, provide protection against radiation damage, lower cholesterol levels and prevention of opportunistic infections in immunocompromised individuals (HIV sufferers). Secondary metabolites, such as halogenated compounds, extracted from macroalgae are of great promise as anti-bacterial and anti-viral agents or as anti-fouling agents. Extracts from certain red algae are used in bone-replacement therapy.

Deep-sea hydrothermal vents now offer a new source of a variety of fascinating microorganisms well adapted to these extreme environments. This new bacterial diversity includes strains able to produce new molecules such as polymers and other bioactive molecules. Screenings of isolates recovered from oceanographic cruises led to the discovery of a number of unusual microbial exopolysaccharides with interesting chemical and rheological properties (Weiss, 1994).

Among these polymers, poly-6-hydroxyalkanoates (PHA) are of special interest. In the same range of high molecular weight biological polymers, chitin, a co-polymer of Nacetylglucosamine and glucosamine and chitosan are found associated with proteins in the exoskeleton of many invertebrate species (annelids and shellfish). Chitin polymers are natural, non-toxic and biodegradable and they have many applications in food and pharmaceuticals as well as cosmetics.

Biomolecular Materials

Recent research has demonstrated that marine biochemical processes can be exploited to produce new biomolecular materials. Due to their diversity, versatility and unusual combinations of properties, biomolecular materials offer promise for application in virtually all sectors of the economy including defence, energy, agriculture, health and environmental technology (Bunker *et al.*, 1994).

For example, a corporation in Chicago is commercialising a new class of biodegradable polymers modelled on natural substances that form the organic matrices of mollusc shells. Equally exciting are the mechanisms used by marine diatoms, coccolithophorids and other marine invertebrates to generate elaborate mineralised structures on a nanometre scale.

Research that will enhance understanding and allow engineering of the processes for creating these bioceramics promises to revolutionise the manufacturing of medical implants, automotive parts, electronic devices, protective coatings and other novel products. Biomaterials also hold promise for counteracting biofouling, which long has been recognised as an extensive and costly problem. A product generated by the seagrass *Zostra marina* (eelgrass), for example, is an effective agent for preventing fouling by bacteria, algal spores, and a variety of hard-fouling barnacles and tube worms (Todd *et al.*, 1993). With continued advances in modern biology, molecular genetics and protein engineering, and with rapid improvements in physicochemical characterisation, novel biomolecular materials can be designed and tailor-made to meet specific needs. This in turn, would expand the possibilities for practical applications of these materials.

Biosensors and Bioindicators

Microorganisms provide the basis for development of sophisticated biosensors, bioindicators and diagnostic devices for medicine, aquaculture and environmental biomonitoring (National Academy of Sciences, 1992). One type of biosensor employs the enzymes responsible for bioluminescence.

The *lux* genes, which encode these enzymes, have been cloned from marine bacteria such as *Vibrio fischeri* and transferred successfully to a variety of plants and other bacteria (Wackett, 1994). Another type of biomonitor that holds great promise is the gene probe, which can be used to identify organisms that pose health hazards or may be useful in research. Specific gene probes can be employed to detect human pathogens in seafood and recreational waters; fish pathogens in aquaculture systems; microorganisms capable of mediating desired chemical transformations (ex. toxic chemical degradation, CO_2 assimilation and metal reduction); and specific fish stocks in fish migration and recruitment studies.

Green fluorescent protein, a naturally fluorescent protein first found in the jellyfish *Aequorea victoria* is now widely employed as a sensitive fluorescent molecular "tag" to identify and localise individual proteins within a cell or a sub-set of cells within a tissue and to follow gene expression in various systems (Wackett, 1994). Biosensors are needed for major threats to human or animal welfare. The development of biosensors would need specific information from the genome of marine organisms and would assist bioscreening for toxic or valuable compounds.

Biopesticides

Natural marine products have the potential to replace chemical pesticides and other agents used to maximise crop yields and growth. Current US expenditures for all pesticides amount to $47 billion annually; by the year 2006, biopesticides from marine and other sources are expected to capture an estimated 10 per cent of this market (Burrill, 1983). An example of a marine biopesticide in use today is Padan *TM*, which was developed from a bait worm's toxin known to ancient Japanese fishermen. This natural pesticide has demonstrated activity against larvae of the rice stem borer, the rice plant skipper and the citrus leaf miner.

More recently, scientists discovered novel compounds in marine algae and marine sponges containing symbiotic microorganisms. These compounds promoted growth and stimulated germination and increased root and coleoptile lengths in test plants. Several sponge and nudibranch species produce terpenes, a broad class of aromatic compounds used in solvents and perfumes and known to deter feeding by fish (Cardellina, 1986). Extracts derived from these same sponges and nudibranch species also demonstrated powerful insecticidal activity against two species, grasshoppers and the tobacco hornworm.

Biorefining

Biorefining involves the use of microbes in mineral processing systems. Biorefining is environmentally friendly and in some cases enables the recovery of minerals and use of resources that otherwise would not be possible. For example, a significant amount of copper is biorefined from slag heaps (Fenical, 1993). Research now under way or planned addresses the use of microorganisms to bioleach oxide and sulphide ores and to concentrate metals. Current research on bioleaching of oxide and sulphide ores addresses treatment of manganese, nickel, cobalt and precious metal ores. Use of microorganisms to remove fine particles from aqueous process streams is also of interest. Laboratory studies have shown that certain microorganisms can cause flocculation of fine mineral suspensions, while others can function as flotation collectors or depressants. Extensive research will be needed before the use of extracellular colloidal molecules can become a viable alternative to current chemical based systems.

Bioremediation

Due to its comparatively low cost and generally benign environmental impact, bioremediation offers an attractive alternative or supplement to more conventional clean-up technologies (Gibson and Sayler, 1992). Bioremediation shows great promise for addressing problems in marine environments. These problems include catastrophic spills of oil in harbours and around oil platforms; movement of toxic chemicals from land through estuaries into the coastal oceans; disposal of sewage sludge and chemical process wastes; reclamation of minerals, such as manganese; and management of aquaculture and seafood processing waste (Russell, 1992). The full potential for marine organisms and processes to contribute new waste treatment and site remediation technologies cannot be realised without enhanced understanding of the unique conditions in marine environments.

Recent developments in biology have provided new tools and approaches for monitoring the environment and engineering organisms with the capacity to degrade environmental pollutants. These developments have created unprecedented opportunities for significant advances (Miller and Poindexter, 1994).

Indeed, bioremediation is expected to become an industry with annual sales of more than $500 million by the year 2010 (National Research Council, 1993). Investment in bioremediation research has the dual benefits of solving important environmental problems while stimulating the growth of the bioremediation industry. Recently, a technique, which encourages the growth of resident bacteria through fertilization for the clearing of oil spills, was developed in temperate zones.

Research continues to verify the bioremediation potential

of microorganisms. For example, a recent addition to the growing list of bacteria that can sequester or reduce metals is *Geobacter metallireducens*, which removes uranium, a radioactive waste, from drainage waters in mining operations and from contaminated groundwater (US-EC Task Force on Biotechnology Research, 1992).

Bioelectronics and Bionetworks

Bioelectronics is an emerging technology that employs biological molecules instead of inorganic materials in conventional integrated-circuit technology or in applications involving unconventional architectures, such as optical processors (Zilinskas *et al.*, 1994). The driving force for this research is the possibility of constructing devices on the molecular level and thereby achieving extremely high densities of data storage sites and nano-sized computers.

Biological systems are capable of storing information on the molecular level and processing information along pathways defined on a molecular level. Although biological processes function more slowly than do conventional solid-state devices, this penalty is more then offset by a huge increase in the density of operating units. Various biomimetic or biologically based materials, such as the protein bacteriorhodopsin, are being evaluated for use in bioelectronics (Burrill and Lee, 1993). Recent achievements in the culturing of a monolayer of neurons on a microelectrode array promise to provide some insight into the operation of neuron arrays. Although neuron devices, bionetworks would not operate on a molecular scale, they have the potential to form the basis for new computer architectures, including parallel processors.

Genetic Resources Conservation

The preservation and enhancement of biodiversity in natural systems is an important issue. Biotechnology can be employed in two ways to conserve genetic resources of aquatic species (Langer and Vacanti. 1993). First, the tools of biotechnology can be used to identify and characterise important aquatic germplasm, including endangered species. Genomes of aquatic species can be analysed and characterised, and quantitative trait loci identified. Second, biotechnology can be applied to improve understanding of the molecular basis of gene regulation and expression as well as sex determination and thereby improve methods for defining species, stocks and populations.

Approaches include developing marker-assisted selection technologies, improving efficiency of transgenic techniques, and improving technologies for the cryopreservation of gametes and embryos. Ultimately, stocking certain areas with selected cultivated species and strains could help maintain biodiversity in natural aquatic ecosystems.

Biomass and Energy Production

Biomass includes organic polymeric material such as lignin, starches, celluloses and oils produced by biological processes. Plants and algae are the main sources of biomass, generating many billions of tons annually through photosynthesis. Annually, approximately 40 per cent of all primary energy production (photosynthesis) occurs in the seas (Suttle *et at,* 1990). Biomass is a potential source of energy resources such as methane and ethanol, commodity chemicals, animal feed and speciality products (flavours, fragrances and pigments).

Biomass is an attractive alternative to petroleum based sources because it not only is a low-cost substrate for production, but also is environmentally friendly; it is renewable and may reduce emissions during processing and improve the biodegradability of end products.

Biotechnology is especially relevant to the commercial use of biomass because it employs natural processes uniquely suited for creating specific products (Allen and Friese, 1992). An example of such a process is fermentation, which converts glucose sugars derived from biomass into commodity chemicals. Recently, the marine microalgae are being genetically engineered to boost their lipid content, with the aim of providing a source of alternative fuels that is more economical than are conventional sources. The biotechnology techniques are also useful to convert biomass to ethanol and other alternative forms of energy.

Sustainable Aquaculture and Fisheries: Exploitation of fisheries' resources has increased at a rapid pace and needs to be controlled. Aquaculture can supply high-quality food in a sustainable way. The application of biotechnology promises significant benefits to both producers and consumers of aquacultural products (Economic Research Service, 1994).

The use of genetically enhanced organisms may improve production efficiency through improvements in growth rates, food conversion, disease resistance, and product quality and composition. Biotechnology can be applied to enhance reproduction and early development of cultivated aquatic organisms. The resulting benefits could include year-round production of gametes and fry of economically valuable species and creation of new markets for specialised, genetically improved broodstock. Similarly, biotechnology may provide techniques for improving the reproductive success and survival of endangered species, thereby helping to preserve the diversity of life on earth. Biotechnology not only can improve the survival, growth, vigour and well-being of cultivated stocks, but also can reduce disease transfer between cultivated and wild stocks. Investigations into the use of marine microorganisms as probiotics in aquaculture are resulting in improved fish health and production. Molecular biology techniques can be used to monitor genetic improvements using traditional breeding methods.

Environmental Quality and Management

Marine microorganisms, either as independent strains or as members of microbial consortia, express novel biodegradation pathways for breaking down a wide variety of organic pollutants. Several such consortia have been described but many others should be explored. This will lead in turn to the extensive development and subsequent control of such bioremediation processes in shaping new applications in environmental biotechnology. Moreover, in addition to biodegradation, marine microorganisms frequently produce environmentally friendly chemicals such as biopolymers and non-toxic biosurfactants that can also be applied in environmental waste management and treatment (DeLong, 1992).

In addition, further understanding of the interaction of marine microbes with toxic heavy metals or radionuclides has suggested their use in various bioabsorption, bioprecipitation and biocrystallisation applications for treatment of contaminated water systems. The biosensors have the capacity to operate unattended in automated systems accessible by remote sensing to provide data for large-scale surveillance of marine environments to enable forecasts of major threats to fisheries, aquaculture or public health. More recently, biotechnology has been used to assess seafood safety through the development of novel test kits that react to the presence of human pathogens in the food. Moreover, specific marine bacteria have been developed that can aid in the clean-up effort if a toxic spill should occur; thus biotechnology not only helps society, but it helps the environment as well.

Global Climate Change

Powerful tools are being developed to elucidate the many biogeochemical cycles that determine the fate of all the life supporting elements on earth. Scientists are beginning to understand and manipulate the molecular genetics and biology of esoteric metabolic pathways associated with the carbon, sulphur, phosphorous, iron and other biogeochemical cycles. For example, based on the hypothesis that iron controls photosynthesis in the oceans, immunological probes were used to show that addition of iron to open ocean water off the Galapagos Islands significantly increased energy production (Kolber, 1994). Marine biotechnology will be useful in assessing the role of the oceans in affecting climate change and the global carbon cycle.

Molecular techniques can facilitate and enhance the measurements of CO, concentrations and total CO_2 inventories being developed for global ocean models of carbon cycling. There is compelling evidence that the exchange of dissolved and particulate materials between the continental shelf and its boundaries is a significant factor affecting the flux of CO_2 and biogenic elements within the global ocean.

Potential Areas for Research and Development

Marine biotechnology is an area of key national and international importance particularly in R & D as a foundation for the growth of industry, technology and medicine. The vision for the 21st century is that marine biotechnology will apply advanced tools from molecular biology and information technology to a carefully selected suite of marine habitats and organisms representing the total diversity of marine systems, in order to obtain novel genes and processes that can be turned into new products and approaches for the benefit of industry, biomedicine and sustainable management of the world's oceans (Zilinskas *et al.*, 1994). The major potential areas for research and development in marine biotechnology are:

- Develop a fundamental understanding of the genetic, nutritional and environmental factors that control the production of primary and secondary metabolites in marine organisms, as a basis for developing new and improved products.
- Identify bioactive compounds and determine their mechanisms of action and natural function, to provide models for new lines of selectively active materials for application in medicine and the chemical industry.
- Develop bioremediation strategies for application in the world's coastal oceans, where multiple uses, including wastewater disposal, recreation, aquaculture demand prevention and remediation of pollution; and develop bioprocessing strategies for improving sustainable industrial processes.
- Use the tools of modern biotechnology to improve the health, reproduction, development, growth, and overall well- being of cultivated aquatic organisms; and promote the interdisciplinary development of

environmentally sensitive, sustainable systems that will enable significant commercialisation of aquaculture.

- Improve understanding of microbial physiology, genetics, biochemistry, and ecology in order to provide model systems for research and production systems for commerce, and to contribute to understanding and conservation of the seas.

The areas presented in the following are intended to convey some directions that appear to be important and promising areas for identification and development:

Major fields	*Potential areas*
Genomics and Proteomics	Organismal databases and networks; Novel genes and gene products; Profiling genes and metabolic pathways; Identification of model organisms.
Bioinformatics	New algorithms for specific predictions of drug actions based on models of biological processes.
Nanotechnology	New materials; Feasibility of carrying out biological and chemical processes at the molecular and atomic level.
Bioprospecting	Bioactive molecules; Pharmaceuticals; Added value products from fish waste; Novel drug targets.
Environmental biotechnology	New processes; Biosensors and biomonitoring; Bioremediation; Detection and control of biofouling.
Aquaculture	Environmentally friendly production systems; Health improvement; Improved vaccines and microbial control; Models of host-pathogen interactions; Improvements in productivity of marine organisms; Mariculture of novel species; Production of novel biomolecules; Production of germ-free marine animals.

Contd...

Major fields	*Potential areas*
Cell culture	Basic insights into fundamental cellular processes in marine model organisms; Reliable biopharmaceutical production.

For the last two decades the common conclusions of international fora considering strategic challenges in science have uniformly identified the marine biotope as a large and untapped area for exploration.

The rich diversity of marine form and function, and its unique physiological adaptations to the harsh marine environment coupled with new developments in biotechnology, has opened up a new and exciting vista for the extraction of bioactive products of use in medicine, novel industrial processes and environmental monitoring. Yet to this day no concerted and focused initiative to realise this vision has materialised in developing countries. Current events and needs suggest that now is the time for such action. A common global scientific forum will be needed in marine biotechnology to encourage a new generation of young scientists to contribute their talents and energy to the field and thus further its success. The urgency of the problems in terms of human health and nutrition, environmental impact, and also a new ability to understand and further biodiversity and balanced reproduction, add to the attraction and potential of this initiative. The combined expertise of many fields from biology to chemical and physical oceanography will be needed in order to understand how life and its propagation is shaped by the parameters of the marine biotope. A multi-continent initiative will integrate these activities with clear synergies and enthusiasm stemming from a new focus, and with broad appeal to industry, to biomedicine and to society at large.

Bibliography

Alfoldi, L.: *Hydrogeologic Considerations for in situ Bioremediation,* Butterworth-Heinemann Publishers, Stoneham, 1991.

Anderson, T. A. and Coats, J. R.: *Bioremediation Through Rhizosphere Technology,* American Chemical Society, Washington, 1994.

Antebi, Elizabeth, and David Fishlock: *Biotechnology, Strategies for Life,* Massachusetts, Cambridge, 1986.

Bailey, Ronald: *Ecoscam, the False Prophets of Ecological Apocalypse,* St. Martin's Press, New York, 1993.

Barley, S., J. Freeman and R. Hybels: *Strategic Alliances in Commercial Biotechnology,* Harvard Business School Press, Boston, 1992.

Batra, R. and W. Klassen: *Public Perception of Biotechnology,* Agricultural Research Institute, Maryland, 1987.

Biotechnology Industry Organization (BIO): *Agriculture and the New Biology,* The Association, Washington, 1987.

Biotechnology Research Subcommittee: *Committee in Fundamental Science and Research,* National Science and Technology Council, Washington, 1995.

Blakely, E.: *The Economic Development Potentials of California's Biotech Industry, Biotech Industry Research Group, Institute of Urban and Regional Development,* University of California, Berkeley, 1989.

Blakely, Edward J. and Kelvin W. Willoughby: *Institute of Urban and Regional Development, Biotechnology Industry Research Group,* University of California, Berkeley, 1990.

Blum, Deborah: *The Monkey Wars,* Oxford University Press, London, 1994.

Booz, Allen & Hamilton: *Foreign Investment in Bay Area Bioscience,* Bay Area Bioscience Center, Oakland, 1991.

Center for Agricultural Molecular Biology: *Cook College, New Jersey Agricultural Experiment Station, Rutgers,* The State University, New Brunswick, 1991.

Chong, Ken P. and Smith, John W.: *Mechanics of Oil Shale, London, England and New York,* Elsevier Applied Science Publishers, New York, 1984.

Cirpka, O. and Bisch, G.: *Technical Design of the Aquifer Model and Influence of Spatial Variability,* Presented at the Third International Symposium on In-Situ and On-Site Bioreclamation, California, 1995.

Committee on In Situ Bioremediation Water Science and Technology Board Commission on Engineering and Technical Systems National Research Council: *In Situ Bioremediation, When does it Work,* National Academy Press, Washington, 1993.

Committee on Scientific Evaluation of the Introduction of Genetically Modified Microorganisms and Plants into the Environment: *Field Testing Genetically Modified Organisms, Framework for Decisions,* National Academy Press, Washington, 1989.

Comstock, G.: *What Obligations have Scientists to Transgenic Animals, College Station, Center for Biotechnology Policy and Ethics,* Texas A&M University, *New Jersey,* 1992.

Cooney, J. J. and Gadd, G. M.: *Metals and Microorganisms, Journal of Industrial Microbiology,* Washington, 1995.

Crawford, Ronald L. and Don L.: *Crawford, Bioremediation*

Principles and Applications, Cambridge University Press, New York, 1996.

Dale, Jeremy W.: *Molecular Genetics of Bacteria,* John Wiley and Sons, New York, 1989.

De Lorenzo, V.: *Genetic Engineering Strategies for Environmental Applications,* Current Opinion in Biotechnology, California, 1992.

Drexler, K. Nanosystems: *Molecular Machinery, Manufacturing, and Computation,* John Wiley and Sons, New York, 1992.

Economidis, I. and Palmisano A.: *Environmental Biotechnology, EC-US Task Force on Biotechnology Research,* Brussels, Belgium, 1995.

Ehrlich, H. L.: *Geomicrobiology,* Marcel Dekker, New York, 1990.

Esser, K.: *Plasmids of Eukaryotes, Fundamentals and Applications,* Springer-Verlag Berlin Heidelberg, Germany, 1986.

Fowle, John R.: *Application of Biotechnology, Environmental and Policy Issues,* Westview Press, Colorado, 1987.

Francis, A. J.: *Microbial Dissolution and Stabilization of Toxic Metals and Radionuclides in Mixed Wastes,* Oxford, New York, 1990.

Frankenberger, W. T. and Benson S. M.: *Selenium in the Environment,* Marcel Dekker, New York, 1994.

Friedman, D.: *The New Economy Project Final Report,* The New Economy Team, UCLA, 1994.

Gendel, Steven: *Agricultural Bioethics, Implications of Agricultural Biotechnology,* Iowa State University Press, Ames, 1990.

Gibson, D. T. and Sayler, G. S.: *Scientific Foundations of Bioremediation, Current Status and Future Needs,* American Academy of Microbiology, Washington, 1992.

Ginzburg, Lev. R.: *Assessing Ecological Risks of Biotechnology,* Butterworth-Heinemann, Boston, 1991.

Goodsell, David S.:*The Machinery of Life*, Springer-Verlag, New York, 1993.

Grimes, D. J.: *Validation of Reliability and Predictability of Environmental Biotechnologies*, Workshop on Bioremediation, Japan, 1995.

Grocery Manufacturers of America: *Safe Foods Developed Through Biotechnology, Nutritious, Healthful,* Abundant and Good Tasting, 1993.

Hacking, Andrew J.: *Economic Aspects of Biotechnology,* Cambridge University Press, Cambridge, 1986.

Harkness, M. R., McDermott J. B. and Abramowicz D. A.: *In Situ Stimulation of Aerobic Degradation of PCB Biodegradation in Hudson River Sediments,* MacMillan, London, 1993.

Hettinger, N.: *Owning Varieties of Life, Biotechnology, Intellectual Property, and Environmental Ethics, Center for Biotechnology Policy and Ethics,* Texas A&M University, Texas, 1994.

Hettinger, N.: *Biotechnology Policy and Ethics,* A&M University, Texas, 1992.

Hinchee, R. E., Anderson D. B., Metting, F. B. and Sayles G. D.: *Applied Biotechnology for Site Remediation,* Lewis Publishers, Florida, 1994.

Hinchee, Robert E.: *Applied Biotechnology for Site Remediation,* CRC Press, Ann Arbor, 1994.

Hollenberg, C. P. and H. Sahm: *Microbial Genetic Engineering and Enzyme Technology,* Gustav Fischer, New York, 1987.

Kelly, Kevin: *Out of Control,* Addison-Wesley, Menlo Park, 1994.

Kenney, Martin *Biotechnology: The University-Industrial Complex,* Yale University Press, Connecticut, 1986.

Kimbrell, Andrew: *The Human Body Shop,* Harper/Collins, San Francisco, 1993.

King, R. B., Mihony, G. and Sheldon J. K.: *Practical Environmental Bioremediation,* Lewis Publishers, Florida, 1992.

Kloppenburg, Jack Ralph: *The Political Economy of Plant Biotechnology,* Cambridge University Press, Cambridge, 1988.

Koehler, Gus.: *Tools to Promote Economic Success,* California Research Bureau, Sacramento, 1995.

Kotkin, Joel and Levy, Stephen: *A Twenty-First Century Prospectus, Center for the New West, and Center for the Continuing Study of the California Economy,* California, 1996.

Krimsky, Sheldon: *Biotechnics and Society,The Rise of Industrial Genetics,* Praeger, New York, 1991.

Lacy, W., L. Busch, and L. Lacy: *Agricultural Biotechnology, Issues and Choices,* Purdue University Press, Indiana, 1991.

Lacy, W., L. Lacy and L. Busch: *Emerging Trends, Consequences, and Policy Issues in Agricultural Biotechnology,* Westview Press, Colorado, 1991.

Lee, V. and Yen T.F.: *Biopolymer Shields, in Applied Biotechnology for Site Remediation,* Lewis Publishers, Florida, 1994.

Levorsen, A. I.: *Geology of Petroleum, San Francisco,* W. H. Freeman and Company, California, 1967.

Lovley, D. R.: *Dissimilatory Metal Reduction,* Annual Review of Microbiology, 1993.

Lynch, J. M. and Hobbie J. E.: *Concepts and Applications in Microbial Ecology,* Blackwell Scientific Publications, England, 1988.

McKinsey Global Institute: *Manufacturing Productivity,* McKinsey and Company, Washington, 1993.

Metting, Jr., F. B.: *Soil Microbial Ecology, Applications in Agricultural and Environmental Management,* Marcel Dekker, New York, 1992.

Miller, Henry I., M.D : *Remarks for the California Legislature*

Assembly Committee on Agriculture, MacMillan, London, 1994.

Miller, R. V. and Poindexter J. S.: *Strategies and Mechanisms for Field Research in Environmental Bioremediation,* American Academy of Microbiology, Washington, 1994.

Mueller, J. G.: *Model Studies of Bioremediation Strategies and Monitoring Technologies in Proceedings, Environmental Biotechnology,* EC-US Task Force on Biotechnology Research, Belgium, 1995.

National Research Council (NRC): *Alternatives for Ground Water Cleanup,* National Academy Press, Washington, 1994.

Old, R. W. and S. B. Primrose: *Principles of Gene Manipulation, An Introduction to Genetic Engineering,* University of California Press, California, 1980.

Olson, Steve: *An Industry Comes of Age,* National Academy Press, Washington, 1986.

Organisation for Economic Cooperation and Development (OECD): *Biotechnology for a Clean Environment, Prevention, Detection, Remediation,* OECD, Paris, 1994.

Ouellette, Robert P. and Paul N.: *Cheremisinoff, Essentials of Biotechnology, Lancaster,* Technomic Publishing Company, Pennsylvania, 1985.

Panem, Sandra: *Biotechnology, Implications for Public Policy,* The Brookings Institution, Washington, 1985.

Powell, W. and P. Brantley: *Competitive Cooperation in Biotechnology, Learning Through Networks,* Harvard Business School Press, Boston, 1992.

Prentis, Steve: *Biotechnology, A New Industrial Revolution,* George Braziller, New York, 1984.

President's Council on Competitiveness: *Report on National Biotechnology Policy,* The Council, Washington, 1991.

Rissler, Jane, and Margaret Mellon: *Perils Amidst the Promise,*

Ecological Risks of Transgenic Crops in a Global Market, Union of Concerned Scientists, Cambridge, 1993.

Saaty, R. P. and Booth S. R.: *Cost Effectiveness of a Remediation Technology Field Tested at the Savannah River Integrated Demonstration Site*, Los Alamos National Laboratory, New Mexico, 1994.

Saunders, Venetia A. and Jon R. Saunders: *Microbial Genetics Applied to Biotechnology, Principles & Techniques of Gene Transfer & Manipulation*, Crook Helm Ltd., London, 1987.

Saxenian, A.: *Regional Advantage*, Harvard University Press, Boston, 1994.

Sayler, G. S., Fox, R. and Blackburn, J. W.: *Environmental Biotechnology for Waste Treatment*, Plenum Press, New York, 1991.

Sayler, G. S.: *Biodegradation Process Analysis, Molecular Application in Simulations and Environmental Verification in Proceedings, Environmental Biotechnology*, EC-US Task Force on Biotechnology Research, Belgium, 1995.

Sayler, Gary S.: *Environmental Biotechnology for Waste Treatment*, Plenum Press, New York, 1991.

Skladany, G. J. and Metting, F. B.: *Bioremediation of Contaminated Soil in Soil Microbial Ecology*, Marcel Dekker, New York, 1992.

Smith, George P.: *The New Biology, Law, Ethics and Biotechnology*, Plenum Press, New York, 1989.

Sussman, M.: *The Release of Genetically-engineered Micro-organisms*, Harcourt Brace Jovanovich, New York, 1988.

Thomas, J. M. and Ward, C. H.: *In Situ Biorestoration of Organic Contaminants in the Subsurface*, Environmental Science and Technology, 1989.

Thompson, P: *Animal Welfare and Animal Rights, Center for Biotechnology Polilcy and Ethics*, Texas A&M University, Texas, 1992.

Thompson, Paul B., Robert J. Matthews: *Ethics, Public Policy and Agriculture*, MacMillan, London, 1994.

Tunlid, A. and White, D. C.: *Biochemical Analysis of Biomass, Community Structure, Nutritional Status, and Metabolic Activity of Microbial Communities in Soil in Soil Biochemistry*, Marcel Dekker, New York, 1992.

Wijesinghe, A. M., Knapp, R. B. and Carman, L. M.: *Preliminary Feasibility and Cost Analysis of the In-Situ Microbial Filter Concept*, Lawrence Livermore National Laboratory, Livermore, California, 1992.

Index

A

Academia, 166, 173, 174.

Acid, 25, 40, 42, 43, 44, 45, 46, 58, 85, 100, 122, 206, 219, 220, 221, 222, 223, 224, 225, 226, 227, 228, 229, 252, 269, 294.

Agriculture, 22, 52, 55, 72, 104, 126, 129, 130, 131, 137, 138, 139, 140, 145, 146, 148, 158, 166, 167, 170, 172, 176, 177, 178, 181, 187, 188, 232, 233, 235, 237, 238, 239, 240, 242, 246, 255, 256, 262, 264, 272, 276, 283, 287, 298.

Amino Acids, 130, 142, 222, 225, 228, 257, 297.

Animals, 2, 3, 7, 8, 9, 11, 12, 18, 19, 20, 22, 23, 24, 26, 31, 32, 34, 47, 48, 51, 52, 58, 61, 85, 87, 90, 91, 94, 95, 96, 98, 99, 102, 103, 107, 117, 124, 125, 131, 139, 140, 142, 148, 150, 151, 152, 154, 182, 183, 186, 232, 238, 243, 256, 258, 264, 265, 280, 295, 307.

Aquaculture, 73, 74, 85, 102, 103, 108, 112, 114, 117, 118, 120, 124, 156, 243, 297, 299, 301, 304, 305, 306, 307.

Arsenic, 42, 49, 59.

B

Bacteria, 3, 5, 7, 12, 34, 35, 90, 100, 112, 115,

117, 129, 130, 142, 199, 200, 202, 222, 290, 291, 294, 295, 296, 298, 299, 301, 302, 305.

Biochemical, 10, 106, 109, 114, 117, 199, 201, 203, 237, 297, 298.

Biodiversity, 8, 19, 20, 28, 72, 73, 126, 140, 148, 151, 156, 233, 236, 237, 238, 240, 245, 266, 275, 286, 302, 303, 308.

Biofertilizers, 137, 138, 142.

Biology, 52, 88, 97, 99, 104, 111, 112, 115, 118, 121, 139, 142, 190, 247, 255, 270, 298, 301, 304, 305, 306, 308.

Biomass, 9, 11, 17, 31, 32, 142, 203, 208, 213, 214, 215, 216, 217, 218, 219, 220, 225, 226, 227, 229, 259, 303, 304.

Biomedical Research, 85, 86, 97, 102.

Biomolecular Materials, 298, 299.

Bioremediation Processes, 290, 304.

Biosensors, 299, 305, 307.

Biotechnology, 52, 103, 109, 121, 123, 124, 125, 127, 128, 129, 131, 136, 138, 139, 140, 145, 146, 147, 148, 157, 165, 166, 167, 168, 170, 171, 172, 173, 174, 175, 176, 179, 180, 181, 182, 183, 184, 185, 186, 187, 188, 189, 190, 191, 192, 193, 194, 224, 232, 233, 235, 237, 238, 240, 241, 242, 243, 244, 245, 246, 247, 248, 253, 255, 256, 257, 259, 260, 261, 263, 264, 265, 266, 268, 269, 270, 271, 272, 273, 274, 275, 276, 277, 278, 279, 280, 281, 283, 284, 285, 286, 287, 288, 289, 290, 291, 292, 293, 294, 295, 302, 303, 304, 305, 306, 307, 308.

C

Cadmium, 49, 59, 65.

Calcium, 34, 79, 225.

Carbon, 2, 3, 4, 10, 11, 12, 13, 22, 27, 32, 40, 112, 221, 226, 305, 306.

Carbon Dioxide, 2, 3, 12, 13, 22, 27, 40, 112.

Chemical, 4, 39, 40, 46, 47, 48, 52, 54, 55, 56, 58, 67, 75, 86, 87, 88, 89, 90, 91, 92, 93, 96, 133, 134, 136, 147, 149, 155, 179, 188, 190, 191, 193, 197, 207, 222, 229, 260, 289, 293, 295, 296, 297, 299, 301, 306, 307, 308.

Chlordecone, 55, 62.

Chromium, 45, 49, 59.

Climate, 19, 20, 22, 27, 28, 29, 30, 232, 291, 305.

Coffee Tissue Culture, 249, 250.

Copper, 40, 59, 109, 213, 214, 215, 216, 217, 218, 219, 220, 270, 300.

Corn, 112, 132, 133, 134, 135, 136, 168, 263.

D

Dichromate, 45, 46.

Dioxins, 58, 59.

Disease, 17, 47, 53, 72, 87, 88, 95, 99, 104, 106, 111, 118, 123, 126, 127, 155, 162, 170, 243, 249, 250, 294, 304.

Diversity, 7, 14, 21, 22, 23, 24, 27, 30, 31, 73, 85, 86, 94, 111, 120, 126, 139, 149, 153, 170, 171, 187, 232, 233, 234, 236, 237, 238, 239, 240, 266, 279, 283, 290, 291, 293, 297, 298, 304, 306, 308.

Drugs, 20, 27, 47, 104, 116, 124, 143, 243, 290, 294.

E

Ecology, 1, 2, 124, 174, 272, 307.

Embryonic Cells, 71, 83.

Energy, 1, 2, 4, 5, 7, 8, 9, 10, 11, 13, 14, 15, 17, 27, 28, 35, 36, 37, 38, 39, 65, 115, 188, 189, 191, 257, 259, 272, 290, 298, 303, 304, 305, 308.

Environmental Biotechnology, 109, 129, 188, 192, 194, 289, 304.

Enzymes, 51, 54, 55, 62, 64, 114, 124, 142, 188, 192, 193, 204, 259, 260, 289, 290, 295, 296, 299.

F

Farmers, 13, 74, 131, 132, 133, 134, 135, 137, 138, 139, 146, 147, 148, 149, 150, 151, 153, 154, 157, 158, 159, 160, 161, 163, 164, 165, 167, 170, 174, 176, 177, 178, 179, 180, 232, 233, 235, 240, 242, 249, 250, 269, 281, 285.

Fibre, 49, 131, 257.

Fish, 3, 6, 9, 26, 33, 51, 52, 53, 56, 57, 58, 59, 60, 61, 62, 63, 64, 65, 66, 67, 68, 69, 70, 71, 72, 73, 74, 75, 77, 78, 79, 80, 82, 83, 84, 85, 86, 87, 88, 89, 90, 92, 93, 94, 95, 96, 97, 98, 99, 100, 101, 102, 103, 104, 105, 106, 107, 108, 109, 110, 111, 112, 113, 114, 115, 116, 117, 118, 119, 120, 121, 122, 123, 124, 125, 126, 127, 128, 150, 152, 155, 156, 258, 294, 297, 299, 300, 304, 307.

Food, 2, 3, 5, 6, 7, 8, 9, 10, 11, 12, 15, 19, 27, 28, 48, 55, 63, 85, 86, 100, 101, 102, 103, 108, 117, 124, 125, 131, 137, 138, 139, 140, 141, 143, 146, 166, 167, 170, 171, 175, 176, 188, 213, 220, 232, 233, 235, 236, 239, 240, 242, 245, 254, 256, 259, 264, 267, 271, 278, 283, 284, 285, 288, 295, 296, 297, 298, 304, 305.

Food Industry, 138, 167, 297.

Food Technology, 259, 283.

Food-safety, 267, 283.

Forests, 1, 4, 14, 18, 19, 20, 21, 22, 31, 179.

Fossil Fuels, 12, 188, 259, 290.

Fungi, 3, 5, 7, 20, 22, 117, 130, 142, 198, 200, 207, 235, 251, 294.

G

Gene, 52, 59, 60, 66, 68, 73, 74, 77, 84, 87, 89, 90, 93, 94, 97, 98, 99, 100, 101, 103, 104, 107, 108, 110, 111, 116, 120, 121, 122, 123, 124, 127, 132, 139, 148, 149,

150, 151, 152, 154, 158, 178, 199, 200, 201, 202, 203, 204, 208, 209, 210, 221, 246, 247, 258, 259, 261, 266, 267, 279, 292, 299, 303, 307.

Gene Transfer, 93, 94, 124, 139, 266.

Genetic, 73, 85, 86, 91, 94, 95, 97, 98, 99, 102, 103, 107, 111, 114, 118, 120, 121, 125, 126, 131, 139, 141, 148, 149, 150, 152, 153, 154, 156, 157, 167, 179, 171, 177, 180, 181, 197, 198, 230, 231, 232, 233, 234, 235, 236, 237, 238, 239, 240, 242, 243, 244, 250, 251, 252, 253, 254, 255, 256, 257, 258, 259, 260, 261, 262, 264, 265, 266, 268, 269, 270, 271, 272, 273, 274, 275, 277, 279, 280, 281, 283, 284, 285, 286, 287, 288, 289, 290, 291, 292, 293, 296, 302, 304, 306.

Genomic, 90, 93, 106, 113, 118, 119, 199, 200, 201, 202, 203, 205, 292.

Glucose, 101, 115, 222, 223, 224, 225, 226, 228, 229, 303.

Grassland, 6, 8, 31.

Green Plants, 2, 6, 10, 11, 12, 13, 15, 26, 296.

Groundwater, 24, 302.

H

Health, 48, 49, 51, 52, 54, 59, 70, 87, 95, 96, 99, 100, 111, 114, 118, 139, 140, 142, 143, 151, 164, 167, 170, 171, 186, 190, 213, 237, 245, 266, 284, 295, 297, 298, 299, 304, 305, 306, 307, 308.

Hormones, 51, 53, 54, 55, 56, 61, 62, 64, 143, 250, 258.

Human Health, 48, 49, 70, 100, 308.

Hydrogen, 3, 4, 11, 35, 38, 42, 43, 45, 46, 224.

I

Industry, 46, 55, 58, 72, 124, 138, 156, 166, 167, 181, 184, 185, 186, 187, 190, 192, 193, 213, 229, 230, 233, 235, 236, 240,

241, 244, 247, 272, 275, 288, 289, 290, 295, 296, 297, 301, 306, 308.

Iron, 27, 59, 81, 139, 305.

J

Jellyfish, 107, 121, 294, 299.

Journalists, 147.

K

Kanamycin, 204.

Kinetics, 218, 220, 222, 225, 227, 229.

L

Laboratory, 47, 48, 57, 58, 70, 93, 94, 95, 107, 123, 200, 202, 205, 300.

Labour, 84, 133, 157, 158, 160, 164, 165, 176, 261.

Lakes, 1, 13, 23, 24, 25, 26, 32, 33, 62, 72, 74.

Lead, 48, 55, 59, 62, 65, 69, 71, 77, 120, 131, 139, 140, 145, 184, 237, 241, 256, 260, 263, 284, 286, 294, 304.

M

Magnesium Ions, 79.

Marine Biotechnology, 52, 291, 292, 293, 294, 305, 306, 308.

Medicine, 102, 107, 243, 246, 260, 296, 299, 306, 308.

Mercury, 59, 196.

Metal, 76, 89, 109, 110, 127, 188, 194, 195, 207, 208, 213, 214, 215, 216, 218, 219, 220, 225, 299, 300.

Methoxychlor, 55, 58, 59, 62.

Methyl, 49, 58, 110, 127, 223, 226.

Microbial, 7, 12, 80, 123, 142, 198, 207, 222, 225, 290, 297, 304, 307.

Microorganisms, 2, 3, 5, 8, 12, 13, 108, 115, 118, 131, 207, 237, 241, 259, 260, 261, 280, 289, 290, 291, 294, 296, 297, 299, 300, 302, 304, 305.

Molecular, 12, 53, 75, 77, 95, 99, 102, 104, 111, 115, 116, 118, 123, 126, 139, 209, 236, 247, 255, 261, 270, 292, 298, 299, 302,

303, 304, 305, 306, 307.

Mosquito Fish, 119, 125.

Mountains, 1, 29.

N

Natural Gas, 27.

Nickel, 49, 214, 300.

Nitrogen, 4, 10, 11, 12, 25, 77, 81, 129, 130, 142, 191, 209, 223, 224, 237, 261.

Nuclear, 116, 122, 130, 183, 197.

Nucleus, 120, 127, 197.

Nutrition, 3, 17, 112, 256, 284, 308.

O

Oceans, 1, 13, 14, 27, 28, 29, 214, 290, 291, 292, 293, 301, 305, 306.

Oil, 27, 58, 65, 72, 139, 191, 206, 242, 257, 269, 297, 301.

Oocytes, 62, 74, 82, 83, 113, 123.

Organic, 3, 4, 5, 6, 7, 12, 13, 15, 18, 22, 26, 30, 31, 33, 34, 79, 130, 137, 167, 170, 207, 242, 298, 303, 304.

Organism, 5, 6, 13, 54, 55, 57, 67, 68, 86, 98, 103, 104, 109, 112, 117, 120, 150, 155, 157, 221, 222, 258.

Organochlorine, 52, 58, 59, 60, 62.

Organophosphate, 62, 66, 110.

Oxygen, 4, 11, 26, 27, 33, 42, 74, 120, 223, 224.

P

Pectin Methylesterase, 198, 199, 200, 201.

Phenol, 58, 202, 207, 224.

Phenolic Oxidation, 251, 252.

Phosphate, 13, 223.

Phosphorus, 4, 10, 11, 12, 13, 25, 27, 110, 127, 130.

Photosynthesis, 2, 12, 13, 15, 20, 22, 23, 33, 259, 296, 303, 305.

Plant, 3, 5, 7, 15, 24, 30, 31, 32, 101, 130, 131, 133, 142, 148, 150, 156, 170, 177, 194, 195, 197, 198, 199, 200, 201, 204, 209, 210, 211, 212, 234, 235, 237, 239, 241, 250, 251, 253, 254, 256, 257, 259, 260,

261, 262, 263, 272, 274, 277, 278, 279, 280, 281, 282, 283, 300.

Plastics, 52, 55.

Polymers, 297, 298.

Potassium, 4, 44, 79.

Protein, 63, 67, 97, 101, 102, 104, 105, 107, 108, 109, 112, 113, 114, 115, 116, 117, 121, 122, 127, 132, 198, 199, 200, 201, 202, 205, 228, 244, 297, 298, 299, 302.

R

Research, 53, 56, 57, 61, 69, 70, 85, 86, 88, 89, 92, 93, 95, 97, 101, 102, 103, 104, 107, 108, 112, 113, 117, 118, 119, 121, 124, 125, 126, 128, 131, 135, 137, 139, 141, 142, 145, 147, 152, 154, 159, 165, 166, 168, 169, 170, 171, 172, 174, 175, 176, 179, 181, 187, 190, 209, 213, 230, 231, 233, 236, 239, 241, 242, 243, 244, 245, 246, 247, 248, 249, 253, 255, 258, 260, 261, 263, 268, 269, 273, 275, 276, 277, 278, 279, 281, 282, 285, 286, 287, 288, 289, 290, 291, 292, 293, 295, 296, 297, 298, 299, 300, 301, 302, 304, 306, 307.

Rivers, 1, 24, 26, 27, 87.

S

Salt, 25, 26, 37, 44, 72, 79, 214, 278, 295.

Sea, 1, 7, 11, 13, 26, 27, 33, 34, 72, 81, 101, 108, 291, 294, 295, 297.

Selenium Metabolism, 110, 127.

Sodium, 42, 79, 222.

Soil, 4, 16, 18, 21, 23, 24, 30, 31, 32, 129, 130, 132, 136, 147, 160, 178, 191, 264, 271, 278.

Species, 2, 7, 9, 14, 17, 18, 20, 21, 23, 24, 27, 28, 30, 31, 32, 42, 43, 52, 63, 66, 67, 71, 72, 73, 74, 75, 77, 78, 79, 80, 81, 82, 84, 85, 86, 88, 89, 91, 93, 95, 100, 101, 102, 105, 107, 108, 109, 111,

113, 114, 118, 119, 120, 124, 126, 129, 139, 142, 151, 152, 156, 183, 207, 221, 232, 233, 234, 235, 242, 251, 256, 257, 258, 261, 262, 263, 267, 278, 280, 293, 295, 298, 300, 302, 303, 304, 307.

Sulphide, 3, 58, 300.

Sulphur, 3, 4, 11, 305.

T

Technology, 59, 68, 86, 88, 95, 97, 100, 107, 108, 111, 120, 121, 123, 127, 134, 136, 137, 147, 148, 149, 150, 151, 152, 157, 161, 162, 164, 165, 166, 167, 172, 174, 176, 177, 182, 184, 186, 187, 190, 191, 192, 193, 195, 197, 222, 225, 231, 232, 233, 234, 235, 236, 237, 238, 239, 240, 241, 242, 243, 254, 255, 256, 259, 262, 264, 265, 267, 268, 269, 270, 271, 272, 273, 274, 275, 276, 278, 279, 281, 282, 283, 284, 285, 286, 287, 288, 289, 292, 298, 302, 306.

Temperature, 4, 14, 19, 20, 21, 22, 25, 26, 28, 29, 33, 34, 35, 56, 62, 76, 77, 80, 81, 82, 112, 196, 208, 210, 212, 222, 223, 224, 292.

Terminator Technology, 148, 176, 177, 235.

Thiourea, 58.

Tissue, 59, 64, 90, 91, 94, 98, 107, 120, 122, 127, 139, 147, 195, 198, 199, 200, 201, 208, 209, 210, 211, 212, 233, 234, 242, 244, 246, 248, 249, 250, 251, 261, 269, 277, 295, 299.

Tobacco, 47, 132, 199, 200, 201, 202, 204, 205, 262, 263, 300.

Toxicology, 85, 86, 88, 93, 95, 96, 103, 109, 112, 119, 126.

Transgenic Fish, 86, 87, 88, 89, 90, 92, 93, 94, 95, 96, 97, 99, 100, 101, 108, 120, 121, 122, 124, 125, 152.

Transgenic Plants, 148, 200, 202, 204, 205, 206, 254, 262, 263, 266, 270.

U

Urea, 58, 130.

V

Vanadium, 45.
Various Ecosystems, 16.
Virus, 100, 105, 106, 116, 127, 132, 133, 199, 201, 234, 258, 262, 263, 264, 294.
Vitamins, 12, 142, 222, 225, 228, 257, 297.
Voltage, 37, 38, 196.
Vunisa Cotton, 158, 159, 161, 164.

W

Water, 2, 4, 6, 11, 13, 14, 17, 18, 19, 23, 24, 26, 27, 28, 29, 30, 32, 34, 36, 40, 48, 52, 55, 58, 61, 62, 63, 69, 70, 72, 73, 75, 76, 82, 85, 87, 89, 97, 110, 117, 119, 129, 136, 164, 191, 195, 196, 197, 198, 209, 212, 213, 214, 216, 260, 264, 271, 272, 290, 305.
Wood, 20, 72, 207.

Z

Zebrafish, 52, 60, 65, 90, 93, 97, 98, 99, 104, 105, 107, 110, 117, 119, 121, 122, 123, 124, 127.
Zinc, 59.

❑❑❑